I0084938

KILLER
COMPANY

MATT PEACOCK
KILLER COMPANY

THE BOOK THAT INSPIRED THE ABC1 SERIES DEVIL'S DUST

ABC
Books

The ABC 'Wave' device is a trademark of the Australian Broadcasting Corporation and is used under licence by HarperCollins*Publishers* Australia.

First published in Australia in 2009
This edition published in 2012
by HarperCollins*Publishers* Australia Pty Limited
ABN 36 009 913 517
harpercollins.com.au

Copyright © Matt Peacock 2009

The right of Matt Peacock to be identified as the author of this work has been asserted by him in accordance with the *Copyright Amendment (Moral Rights) Act 2000*.

This work is copyright. Apart from any use as permitted under the *Copyright Act 1968*, no part may be reproduced, copied, scanned, stored in a retrieval system, recorded, or transmitted, in any form or by any means, without the prior written permission of the publisher.

HarperCollins*Publishers*
Level 13, 201 Elizabeth Street, Sydney NSW 2000, Australia
31 View Road, Glenfield, Auckland 0627, New Zealand
A 53, Sector 57, Noida, UP, India
77–85 Fulham Palace Road, London W6 8JB, United Kingdom
2 Bloor Street East, 20th floor, Toronto, Ontario M4W 1A8, Canada
10 East 53rd Street, New York NY 10022, USA

National Library of Australia Cataloguing-in-Publication data:

Peacock, Matt.
  Killer company : James Hardie exposed
  Matthew Peacock.
  ISBN: 978 0 7333 3063 6 (pbk.)
  Banton, Bernie (Bernard Douglas) 1946-2007.
  James Hardie Industries – History.
  Asbestosis – Australia.
  Political activists – Australia – Biography.
  Labor union members – Australia – Biography.
  Asbestos industry – Health aspects – Australia.
  Worker's compensation – Australia – Cases studies.
  Australian Broadcasting Corporation.
616.2440994

Cover photograph by Rob Homer/Fairfaxphotos
Cover design by Matt Stanton
Author photograph by Mark Rogers
Typeset in 11/18pt Sabon by Kirby Jones

Contents

This book is dedicated to all the men and women who have suffered as a result of asbestos exposure, some of whom became good friends while I researched and wrote this book. It is also dedicated to the trade unionists whose often lonely fight has slowly restricted the use of asbestos in many parts of the world.

one

BURYING 'AN AUSTRALIAN HERO'

Bernie Banton lay in his hospital bed. He could hardly talk, but his eyes spoke volumes. This was the pain he'd feared for a decade. It was killing him, and he knew it.

'This is no good, Bernie,' I said gently. 'We still have things to do.'

'I know,' he whispered.

'I'll keep going, though,' I said. 'That's a promise.'

He gave me a thumbs-up.

Across the Sydney skyline from the hospital, the business of Bernie Banton's death was being argued out in a compensation court with representatives of his former employer, James Hardie, once Australia's asbestos giant. This was the second time Banton had sued Hardie in the NSW Dust Diseases Tribunal. He had already successfully claimed compensation in 2000 for the asbestos-induced fibrosis in his lungs that required a regular oxygen supply.[1] This time, the hearing was for full-blown cancer, and had been expedited because of his rapidly deteriorating health.[2] If he died before judgement, most of his claim died with him.

Journalists were crammed into the tribunal courtroom. Some had been there three years before, covering a Special

Commission of Inquiry into James Hardie after it had moved its corporate headquarters to the Netherlands, leaving behind a near-bankrupt foundation to compensate its victims. In the battle to drag the company back and make it pay, Banton had become the face of asbestos sufferers across the country. With his trademark oxygen tubes and tank, he was a household name. Now on his deathbed, public interest in his plight was intense.

Banton's lawyer, Jack Rush, QC, one of the country's most experienced asbestos barristers, rose to his feet before the tribunal's president, Judge John O'Meally, and began tabling a series of Hardie papers.

One piece of correspondence from 1966 was, he said, 'highly significant'.[3] As he explained to me later, the tribunal had accepted it as evidence more than a decade before, so it was not new. But Rush believed it had never been given the attention it deserved because it involved John Reid, whose reign as chairman of James Hardie had lasted twenty-three years.

Rush paused to look over to Hardie's lawyer then turned back to the judge. 'Your Honour, we have called on our learned friends to admit that Mr Reid is still alive,' he said.

'Is he?' asked the judge.

The exchange was barely audible, noticed only by the few journalists straining to hear, but for those who knew Hardie's history this was a moment of high drama. It meant that after more than a quarter of a century of silence about the damage his asbestos company had inflicted on thousands of unwary victims, Reid might finally be called to public account. Hardie's counsel told the court he would seek instructions.

John Boyd Reid, AO, was indeed still very much alive. Now a multimillionaire philanthropist and patron of the arts, he lived only a few kilometres away in a luxury penthouse at Pyrmont Point looking north over Sydney Harbour. Reid had become deputy chairman of James Hardie Asbestos soon after Banton joined the company in 1968, and five years later took over from his father as chairman of the family company. Yet despite his long tenure at the helm, Reid had never appeared in court to answer questions about his role as head of a company responsible for poisoning an estimated 20,000 Australians.[4]

The papers that Rush felt were so important included a query from Reid, then a James Hardie director, written in February 1966, seeking advice from Hardie's personnel manager, Ted Pysden. Reid had read a press clipping from Britain's *Sunday Times* headlined URGENT PROBE INTO 'NEW' KILLER DUST DISEASE sent to him by a colleague, who was concerned about '... this sort of information getting around Australia'. Reid wanted Pysden to advise him on how to respond. According to the *Sunday Times*:

> A disquieting new occupational disease capable of killing not only exposed workmen but also perhaps his womenfolk and even people living near his place of work is the subject of intense behind-the-scenes activity by British scientists, experts in industrial health and at least two government ministries. The condition is a rapidly fatal tumour named mesothelioma that spreads over the pleural covering of the lung. So rare a disorder was it until recent years that many pathologists dispute its existence, but now the growth is being met in increasing numbers at autopsy, especially in patients at one time exposed to asbestos.[5]

To a hushed court, Rush read out the reply to Reid by Hardie's personnel manager:

> The article is not new — it is merely one of many reports on world studies which have been conducted since 1935 when the association between exposure to dust and carcinoma of the lung, mesothelioma of the pleura, tumour of the bladder and uterus and other fatal complaints, was first recognised. The nucleus is dust particles — fibre.

Rush again looked over to the Hardie lawyer, then continued to read from Pysden's response:

> The *Financial Review*, many State newspapers and a large number of Union papers have all featured reports from time to time. If you look for them, there are about a dozen or so articles a year in Australian newspapers and other publications.
> The best advice you can give your friend is to:
> 1. Ignore the publicity — dust is a fact — denials merely stir up more publicity.
> 2. Do something positive about engineering the dust hazard out of existence.

Rush paused, and looked up to underscore his point: the highest echelons in James Hardie knew of mesothelioma and the dangers of asbestos dust two years before Banton began to work for the company.

Banton's legal team intended to table other damning evidence about Hardie. Banton had worked at Hardie BI, an insulation factory (jointly owned with Hardie's competitor CSR) that supplied insulation for power stations and engine rooms across the land. The factory was managed by Hardie and sat across the railway line from the company's main manufacturing plant at Camellia, near Parramatta in Sydney. Banton's lawyers had discovered a document showing that as far back as 1972 the company doctor, after assessing the incidence of asbestos disease, had written: 'The number of cases at Hardie BI is disproportionately high and reflects the unsatisfactory standards of hygiene that have long existed there and still continue.' By then, Hardie had already identified five mesothelioma cases among its employees. 'It must be anticipated that further cases will occur,' Dr Terry McCullagh had warned.[6]

Banton had worked in the dusty factory until just before its closure in 1974. Even by the company's own standards the conditions at Hardie BI were extraordinarily dangerous. Barry Sheppard, Banton's foreman, gave evidence in his case before the tribunal that the dust was up to four inches thick under their feet.[7] The company had issued a free bottle of milk daily to each worker to ease the irritation it caused to their throats.[8] Sheppard used to pack his bottle in the asbestos dust to keep it cold: occasionally the men would have 'snow fights' with the dust. Like Banton, Sheppard was adamant the company had said nothing to him about the danger of asbestos cancers.

In the tribunal foyer other witnesses awaited their turn, including Greg Combet who, as secretary of the Australian Council of Trade Unions (ACTU), had led exhausting negotiations with Hardie for the previous three years with Banton at his side. Karen Banton, Bernie's wife, knowing that her husband only had days to live, had left his bedside for the first time in weeks and was nervously preparing to give evidence.

But the witnesses were never needed. The Reid memo and the suggestion that the former chairman could be called to give evidence had the desired effect. The Hardie defence team appeared rattled. An adjournment was taken and phone calls were made. The proceedings briefly resumed to hear more evidence, but within an hour James Hardie settled. The terms were kept secret. Five days later Banton died.

Banton had never sought a negotiated settlement for his cancer claim. He wanted a show trial. 'They can offer me a million bucks!' he told me fiercely some months before his hearing. 'I don't care what they offer.' He had hoped to force the people from the top of James Hardie onto the stand to publicly defend their actions — people like John Reid. But the cancer enveloping his abdomen had developed faster than expected.

Of all the cancers, mesothelioma is one of the nastiest. It lies in the body for up to forty years or more before the thin, sheeted tumour begins its march, rapidly spreading around the pleura lining the lung or, less commonly, around the peritoneum that surrounds the abdomen. Lacking the telltale lumps of other cancers, it often evades detection. Within months it creates a build-up of fluid in the once supple membrane, causing excruciating pain, and inexorably crushing the lungs, heart or bowel.

Banton knew the symptoms well. The spectre had haunted him since he had discovered nine years before that he had asbestos scarring on his lungs. It was then that he also learnt how deadly mesothelioma was. 'That's the fear you live with daily,' he told me at our first meeting. 'You know, with mesothelioma, the average is about 153 days once they tell you that you have it. That's about the end of the line.'

He was speaking in 2004 during his first TV interview,[9] at the start of his campaign against Hardie's decision to move to the Netherlands without leaving behind sufficient funds to compensate its Australian asbestos victims. At the time, Banton was attending the inquiry into Hardie's corporate flight offshore almost every day, as vice-president of the state-based asbestos support group, the Asbestos Diseases Foundation of Australia. He had joined the group in 2000 after he was first diagnosed with asbestos-related fibrosis. As Banton became more involved with the group's work he had learnt more about the scale of the asbestos tragedy both in Australia and around the world. One third of the world's nations, mainly the poorer ones, still used asbestos and the industry was expanding.[10] Asbestos was still legal in the US and Canada. Indeed, it had only just been outlawed in Australia.[11]

Banton took to television like a duck to water. The unfortunate irony of his condition was that he could only speak in short sound bites. He was already 60 per cent 'dusted', meaning that asbestos scarring had more than halved his lung capacity. By the end of our interview for ABC TV's *7.30 Report*, it was clear I didn't need to include anyone else from the group. 'No offence,' I told his colleague, 'but Bernie here is all you need.' To use an industry phrase, he was ideal 'talent'.

I was no stranger to the dangers of asbestos. Nearly three decades before, as a young reporter, I had made a series of radio programs about the Australian industry for the ABC.[12] I had met many asbestos workers, some of whom had fallen sick and died painful deaths. I had witnessed the disbelief and anger felt by their shattered families as they, and I, gradually learnt more about James Hardie's asbestos cover-up.

Banton and I began to swap information about the company and his campaign. He became a media celebrity, appearing regularly on the nation's top-rating talk shows and television programs. A lunch with Bernie and Karen in the city would be constantly interrupted by well-wishers wanting to shake his hand and, as we made our way up the street at the slow, breathless pace common to former asbestos workers, strangers would continue to accost him. It struck me then that he just might win the campaign to call the company to account and pierce the slick Hardie spin machine. Hardie had spent millions on public relations over the years, smoothing away the country's largest industrial disaster. Now Banton was out-spinning the spinners.

His profile and popularity continued to soar. One incident he proudly recounted occurred when his plane arrived late in Adelaide, where he was to lobby for changes to the state's asbestos compensation laws. No shrinking violet, Banton expressed irritation to the flight attendant, explaining that he had some pressing appointments with politicians. She apologised and announced to other passengers that they should remain in their seats after the landing to allow 'an extremely important passenger' to get off first. Of course, once the plane stopped they ignored her and began getting up to gather their bags. Curtly, the flight attendant repeated the instruction, and they resumed their seats. Very slowly, Banton and his associate made their way down the aisle and the resentment evaporated, replaced instead by spontaneous applause. As other passengers clapped, Banton walked past Federal Minister for Immigration Amanda Vanstone in business class, uncomfortably compelled like the others to remain seated.

For all his new-found fame, Banton never forgot the deadly seriousness of his role. He had travelled to South Australia to join local asbestos disease sufferers and the independent politician Nick Xenophon in lobbying the state Labor government for an asbestos compensation regime similar to the one provided by the Dust Diseases Tribunal in NSW. His lawyer, Tanya Segelov from Turner Freeman, recalled how Banton quietly warned the state premier, Mike Rann, that if the law wasn't changed, he would return to stand on the steps of Parliament House every day of the forthcoming election campaign and urge people not to vote for him.[13]

Rann instructed his attorney-general to change the law along the lines already proposed by Xenophon in his private member's bill, but the Labor attorney resisted key elements, including the provision of special compensation for victims who could no longer care for their family. The campaigners turned to the local media for help. The following week the front page of the *Adelaide Advertiser* featured a photo of a local mesothelioma sufferer, Melissa Haylock and her triplets, headlining her plea to members of parliament: DON'T LET ME DIE IN VAIN.[14] The Rann Government gave in and legislation was soon enacted which included the relevant damages.

This steady succession of asbestos victims steeled Banton in his resolve against Hardie. Many, such as his former factory superintendent, Phil Batson, were to die before him. Batson had described the Hardie BI factory as a 'hellhole ... dusty, noisy and dangerous'. He quit before Banton, in 1972, after learning about the dangers of asbestos during his engineering studies, and later told the tribunal: '... Ridiculous though it might now sound, I had a habit of holding my breath as I walked through the factory.'[15] Banton was proud he had helped Batson's young widow and eleven-year-old son obtain additional payments from the Dust Diseases Board.

Banton only worked at the Hardie BI factory for six short years, but that was enough. Typically, jobs with Hardie were a family affair, and the Banton family was no exception. Bernie's brother Edward was a foreman at Hardie BI and got him the job there; another two brothers, Albert and Bruce, also worked for the company. What appeared to be good family fortune soon turned into family tragedy. Aside from Bernie, Edward died from asbestos disease and Albert developed asbestosis. Bruce still lives with the knowledge that he, too, might one day succumb.

While at the Hardie factory, Bernie was married with a young daughter and used to work double shifts to earn extra money. At least once every shift the dust extraction system would block and a thick cloud of asbestos blew back over the machines, covering all who stood there. He said it resembled a snowstorm, and he was constantly covered with a fine white dust on his face, skin, hair and clothes.[16] Like other former James Hardie workers, ex-employees of the Hardie BI insulation factory held annual reunions in Sydney's Parramatta. These get-togethers were shrinking rapidly. Banton told me in 2004 that only nine former workers were still alive out of 137. The number, as always, was difficult to confirm. James Hardie probably knows, but has never said how many died. By the Christmas before Banton's death there was no point in holding another reunion. Instead, Banton and Sheppard were left to drink a few quiet beers together.

After Banton left Hardie BI he worked for several years at a funeral home where he gained first-hand experience of the way the company dealt with compensation claims made against it. A workmate from the asbestos company had contracted mesothelioma and Banton had promised him he would look after arrangements when he died. The day of the funeral arrived. Banton had spoken with the family and personally supervised the preparations. The body lay in the funeral parlour.

'The bloke was all dressed up and ready,' he told me. It was then he learnt that James Hardie had taken out a court order to obtain a biopsy from the corpse to prove beyond doubt that the death had been caused by mesothelioma. The body had to be taken to Westmead Hospital that morning. 'Those frigging mongrels!' he said. 'The more I think about it … we had to move the body out of the van into the hearse! We only just made it! Those stinking mongrels,' he said, shaking his head.

It was an incident with Hardie that Banton never forgot. He had come from a religious family: his father had founded the Foursquare Gospel church in Parramatta, setting a family tradition for Bernie's younger brother, the Reverend Bruce Banton to follow. It was at this church that Bernie, as a recent divorcee, met his third wife Karen, who was attending the funeral of her husband. According to Karen, Bernie was particularly kind and attentive.

Banton was quite shameless, a quality that was invaluable in the tough negotiations with company lawyers. It was never beyond him to make demands of a senior counsel, the NSW premier, or the prime minister, whether for a bound copy of the Special Commission of Inquiry report into Hardie's move offshore, or even for his own state funeral.

One time he called me, saying, 'Matt, you know who's in town?' He paused for theatrical effect.

'No, Bernie, who?'

'Erin Brockovich!' he declared emphatically. Brockovich was the US paralegal — immortalised by Hollywood — who had fought the huge Pacific Gas and Electric Company on behalf of a tiny polluted township in California.

'Yeah?' I was non-committal.

'Well!' he said. 'What do you reckon? *Bernie Banton meets Erin Brockovich*. Sound like a good headline? Do you think you could hook us up?'

I had to laugh. 'I dunno, Bernie, I'll see what I can do.' I realised that it would probably make a good story and called Brockovich's publicist to tell her of Bernie's interest and give her his number. She seemed just as eager as he was to set up a Brockovich–Banton photo opportunity.

Banton was pleased with his reputation for his snappy media one-liners, and he occasionally tried them out on me before he used them. 'Merry Death. Mere-dith. Get it?' he said once, referring to Meredith Hellicar, by then Hardie's Chair. He was soon to encounter her at close quarters. In 2004 he began negotiations on behalf of asbestos victims with Hellicar, a coolly assured businesswoman in an otherwise all-male company hierarchy.

Hellicar was later to describe Banton and others in his group as 'professional victims'.[17] She was explaining to me why she had told Kerry O'Brien, anchor of ABC TV's *7.30 Report*, that she had never met an asbestos victim.[18] Her answer proved to be a seminal moment in the public debate over Hardie's behaviour. She had anticipated the question, she told me, but had chosen to lie. 'I *had* met an asbestos victim ... [but] I'm not about to tell the bloody media I met an asbestos victim. It's none of the media's business. But boy, did I underestimate the impact of that answer!'

So why not say that?

'Because it's very, very private stuff ... Sure, I could have said, "I'm not talking about it" and everyone would have said, "Ah, she never has!" I thought about it, I had expected it. What Kerry meant was, "Have you met a professional victim?" and the answer was, "No".' Hellicar explained that others had also written to her. 'A number of asbestos victims have been terrific to me. They've written wonderful letters to me and have disassociated themselves from the lies that Bernie has told, and basically said, "We know that the company is trying to do the right thing".' Indeed, she added, 'the Reids ... they've got friends who have died.'

Banton, by contrast, wanted to publicise Hardie's negligence. He was determined not to die quietly and he made asbestos an issue in the 2007 federal election. In his final weeks he continued to agitate on behalf of asbestos victims, joining a campaign to obtain federal government funding for a promising mesothelioma chemotherapy drug. Confined to a wheelchair, he attended a demonstration outside the electoral office of Federal Health Minister Tony Abbott, an event that Abbott later dismissed as a 'stunt'. The ensuing public outrage in the heat of the election campaign soon forced Abbott to apologise.

A few weeks later, Banton was at it again with a new colourful phrase. He took public issue with a new conservative minister, Joe Hockey, who had decided that the union movement was 'dying'. Without the unions, Banton responded, James

Hardie workers would have won 'diddly-squat'. Hockey, too, soon felt the need to clarify his statement.

Banton had never expected to get mesothelioma of the abdomen. Three-quarters of mesotheliomas are around the lung and it was there that he expected it to strike. Looking back, it's possible to see that he began to develop symptoms long before anyone realised their significance. Towards the end of May 2007, following a speaking tour for the Asbestos Diseases Foundation to the country NSW towns of Orange and Bathurst, Banton began to feel lethargic. Professor Chris Clarke, a friend and the thoracic specialist who gave evidence at his final tribunal hearing, had been invited to accompany him. The tour was designed to raise local awareness about the dangers of asbestos in the ubiquitous fibro house common to the area. Community concern was high and both men spoke to public meetings of citizens keen to learn more about the dangers. Asbestos insulation had recently killed a manager at an industrial laundry in Orange and one family had been fighting a long-running battle with the Australian Rail Track Corporation to remove asbestos waste dumped in the backyard of an adjoining railway house.[19]

To look at Banton sitting down or talking, Clarke said, most people would get the wrong impression. When Banton walked up a small incline Clarke realised just how dependent he was on his ever-present oxygen tank. 'I had to send a message to get his wife to bring the car down. I wasn't game enough, even though he was on oxygen, to walk any further with him because my skills at resuscitation these days are not good. He really frightened me.'[20]

Even in these quiet country towns, the name James Hardie still stalked Banton. His elder brother Ted had lived in Orange before succumbing to asbestos disease. In Bathurst, Hardie's presence was even stronger. At Mount Panorama, the company used to jet in its executives and valued customers to entertain them during one of its most successful sponsorship events, the Hardie-Ferodo 500. Known later as the Bathurst 1000, this was Australia's premier motor race, and promoted Hardie's asbestos brake pads. Reid had established relations with Bathurst's Charles Sturt University, which overlooked the race track at Mount Panorama, and he had continued to present students with the annual James Hardie Scholarship. However, by the time Banton visited the town, Hardie's name had become so reviled that the award had been quietly renamed the John B. Reid Scholarship.[21]

On his return to Sydney, Banton was feeling progressively more and more unwell, but still could not pinpoint the problem. The cancer inside him was now wrapped tight around his abdomen. In early August his stomach began to swell. 'Within a few days it was like I was having triplets,' he said. 'I could feel the fluid inside of me; it was like a sixteen-pound bowling ball.'

He still didn't realise what was happening. His wife Karen urged him to see the doctor. He resisted for a few days, but the swelling and pain increased. The doctor had him admitted to hospital where a fluid biopsy was taken. The minute Banton saw its colour — blood red — he knew. Two days later, in Concord Hospital, his friend Chris Clarke confirmed his fears. He had rampant malignant mesothelioma.

Despite the devastating news that he was actually dying, Banton remained focused. His hardest task was to tell Karen, who'd gone home to collect his pyjamas and other things he needed. It was a conversation he later described as 'extremely difficult'. The next day he called in his lawyers, Armando Gardiman and Tanya Segelov from Turner Freeman. Segelov had handled Banton's first case, when Hardie had settled his claim for asbestos-induced fibrosis, and she knew how stubborn he could be.

'Bernie was an incredibly difficult client, because he had a figure in his head and he wasn't going to be moved from that figure,' she recalled with affection. Without the disease he had expected to work for at least another twelve years. Instead, his wife Karen was forced to stop work in order to care for him. Luckily, after some tough negotiations, Segelov had won what he wanted. 'He didn't care what the law was. He would look at you and say, "Well, that's not good enough" and expect you to go away and change the law.'[22]

Segelov, too, could be stubborn. A young, pugnacious asbestos litigator, in Banton's first claim she had insisted on a provisional settlement. Usually in law, once a negligence claim against an employer has been resolved, a claimant cannot sue again for the same act of negligence. But a provisional settlement was a special procedure under the rules of the tribunal that allowed claimants to return for more damages if they contracted another condition, common among those with asbestosis who would later develop cancer. Provisional damages were vigorously resisted by defendants like Hardie. 'They would do anything they could, like offer you another $50,000, for you not to settle provisionally,' she told me. But because Banton was only fifty-three and he'd worked at the notorious Hardie BI factory, Segelov was adamant. She feared that he ran a high risk of eventually contracting cancer.

Segelov's strategy proved well founded. On his deathbed, Banton was about to become the first person to take advantage of the tribunal's rule allowing him a second claim. In addition to normal damages, Banton's lawyers also advised him to sue for 'aggravated damages', which could be awarded to compensate for the personal annoyance a company like Hardie might cause Banton during his trial. Further, he applied for 'exemplary damages' because he claimed that Hardie's behaviour had been so appalling it should be punished as a deterrent to others. These three types of damages had never been awarded against Hardie before. It had preferred to settle rather than risk losing. Sick claimants were also deterred from pursuing them because even if they won, they could be faced with the prospect of a drawn-out appeal by the company.

But Banton was delighted by the prospect of suing Hardie again. If the mesothelioma took the average time to progress in his body then, according to his calculations, he still had 151 days before it would squeeze the life out of him. With an expedited hearing, he felt this was his chance to flush out the secrets the company had kept for so long. What's more, his claim could also test whether Hardie's behaviour in potentially stranding its victims when it moved offshore warranted additional punishment. From his hospital bed he called a media conference. This negligence claim before the tribunal would be a trial everyone would know about.

Even in hospital Banton retained his sense of humour. The gleam was still in his eyes, despite his huge belly. His celebrity status meant he was well known around the ward and he enjoyed bragging about the specialists and medical staff who popped in to wish him well. From his window he proudly identified the site where construction would soon commence on the asbestos research institute he had helped bully and cajole politicians into funding, to be posthumously named the Bernie Banton Centre.

As he embarked upon the chemotherapy he knew would probably have little effect, we discussed my progress in tracking down James Hardie's waste dumps, which both of us feared could pose a hazard for an unsuspecting public. Banton had already introduced me to Alan Overton, Hardie's former national logistics manager, who for decades had supervised the trucks coming and going from each factory. 'He knows everything!' Banton had told me excitedly. 'What went where, when. Where the waste was dumped. Everything! And he's promised me that he'll talk to you.' We had gone to see him and he had agreed to meet with me again.

Overton, like Banton, had grown up in Parramatta and was a pillar of the local community. He was president of the Parramatta Leagues Club, Banton's beloved 'Eels', as well as a director of Westmead Hospital and a host of other charitable concerns. But Overton stonewalled at my second meeting with him, explaining he was not in a position to talk because he was still on a company retainer. 'The company has always done the right thing, you know,' he said.

Banton was not happy with this news when I delivered it to him in hospital. 'He knows where it all is, but he won't say. And he's such a "good bloke",' he said sarcastically. 'The fact is, he's hiding the truth.'

During the next few weeks I visited Banton several times. He had moved back home, although he still had to make regular trips to hospital for chemotherapy and to have his abdominal fluid drained, which knocked him around. On one occasion he was feeling too sick to talk to me, but by the following week he'd improved. He was preparing his will when I arrived. Ill though he was, he was still the same feisty Bernie, and Karen, who'd been at his side for the entire ordeal, was almost superhuman in her stoicism.

Each time I spoke to him he would provide a running update on how many litres of fluid had been drained. 'Four and a half litres! Can you believe it?' he'd exclaim. Karen, though, would give me a more tempered appraisal of his health. The clock was ticking.

When Banton's day in court finally arrived it was evident that the compensation foundation set up by Hardie was prepared to fight him. It argued that Banton's claims for aggravated and exemplary damages had already been paid as part of his earlier settlement. But Hardie's legal team had also made a strategic error that infuriated Banton and stunned his lawyers. They argued that one reason why Banton had been paid a large settlement for his asbestos-related fibrosis in 2000 was because he was already significantly disabled. To prove their point they tabled all Banton's medical reports from the previous proceedings. Among them was a specialist medical opinion that Banton's team had never seen before, because the case was settled and Hardie had no obligation to provide it. The opinion came from an expert Hardie had retained to review an X-ray of Banton's lungs taken by the company in 1973 while he worked at its insulation factory. The specialist opinion was that the X-ray showed an ominous thickening of the pleura which indicated 'definite early signs' of asbestos disease. Hardie's Dr McCullagh had screened the X-ray of the twenty-six-year-old Banton's lungs, one of the several thousand he had arranged to be taken of the Hardie workforce. The X-ray had been filed in the company records without comment. Banton was never told by Hardie what it revealed, and he didn't think to ask. He was young and healthy, as far as he knew.

'They could have held on to the report and we never would have seen it,' said Segelov. 'I was reading through it in the office one night, and I thought "Shit!" That's the first time we'd ever seen it. He had the disease in 1973. He then has eighteen more months of gross exposure in the factory and he is never told.'

She could hardly believe that the Hardie lawyers had submitted the report and immediately amended Banton's claim to include this explosive new information.

Judge O'Meally ruled that Banton was eligible to claim aggravated and exemplary damages. Now all he had to do was prove his case. The Hardie lawyers immediately appealed against this ruling.

By now Banton's health had begun to deteriorate dramatically. He had been in and out of hospital with various infections and the build-up of fluid was unrelenting. On one visit he told me he had thought he was going to 'cash in his chips' the previous night. Worried about her client's health, Segelov tried to arrange a settlement conference on the eve of the appeal hearing, but Hardie's lawyers were in no mood to negotiate. The hearing was to go ahead.

There was one sweet irony for Banton. One of the appeal judges was Justice Kim Santow, the same judge who four years earlier, before he had granted the company permission to move offshore, had been reassured by Hardie that it had made ample provision for future asbestos claimants. Banton's spirits soared when he heard the news, though his health was fading fast.

The three appellate judges heard the case quickly and returned from lunch to issue a unanimous ruling against Hardie: there was nothing to stop Banton from claiming the additional damages for his new disease. And he could certainly claim exemplary damages for the company's behaviour in moving offshore.[23]

The case was back in the tribunal the following day. Banton was near death. The tribunal agreed to take his evidence from the hospital bed. Talkback radio was running hot and Banton's battle was now front-page news. A grim-faced Judge O'Meally arrived at Concord Hospital, clerk and stenographer in attendance. Cameras whirred. Upstairs, Banton had been wheeled into a larger room.

'Can I suggest that this factory was not dusty and dirty in the fashion you describe in your first affidavit?' the Hardie lawyer asked him at the bedside hearing.

'No, that's rot,' Banton replied. 'The factory was putrid.'

He was asked by his own counsel about Hardie's move offshore. 'They tried to abdicate their duties. I can never forgive them for what they did.' And what about the X-ray report he'd never seen? 'It's disgraceful I wasn't told.'

Within twenty-four hours, his claim was settled. The amount remained confidential, but his lawyers were certainly satisfied. The green light had now been given for other claimants to follow the same path.

Banton was past celebrating. He wanted to go home to die. I saw him briefly at the weekend and he gave me a wave of encouragement about pursuing the Hardie saga, but he was in great pain.

Segelov summed it up: 'Bernie was a very religious man and he believed he was going to a better place. And you just wanted Bernie to be in that better place where there was no more suffering.'

Three days later, on 27 November 2007, Banton died.

For Jack Rush, Banton's senior counsel, it had been an unusually harrowing case. He told me a week after the funeral that he'd been honoured when Banton called to ask if he'd run the case. 'It greatly affected me, really. So much so, I have a letter to Karen on my desk now that I reckon I have tried to start and finish twenty times. It's just very difficult.'[24]

He paused before giving his view on Hardie. 'In 1972, for the company to know of cases of mesothelioma and not tell these guys and not do anything and have them working in those conditions, inches deep, you know — I get upset *now* talking to you about it. It's appalling. If you look at the history of the company, then jump through to this decade and look at the way in which they tried to escape all this liability, there is a consistency of conduct ... that can only be said to have permeated right through for decades. It's extraordinary.'

Banton's farewell attracted mourners from all over the country. More than 3000 people, including the NSW premier and federal and state ministers, attended a state funeral at the giant stadium in Sydney's Olympic Park, joining the core of Banton's loyal supporters, trade unionists and asbestos activists. On one of the special trains scheduled to run to the stadium, I met people who had travelled for hours to be at the ceremony. One woman was a union delegate from the Newcastle courts. She had never met Banton, but felt so strongly about the injustice he had fought that she had taken a day's leave to pay her respects. Another man in a wheelchair said he just *had* to be there. He had mesothelioma.

As Banton's coffin was slowly carried out into the rain from the stadium, trade unionists mingled with politicians and friends. 'Good on you, Bernie!' Three cheers rang out.

At the service, the new Labor prime minister Kevin Rudd, who had visited Banton during his election campaign and praised him in a victory speech three days before his death, described him as 'an Australian hero' whose passing would leave the nation poorer. His sister Grace told the mourners that Bernie had made public the wrongs visited 'by wicked men and women on unsuspecting workers: companies do not make decisions, people do'. And Banton's twelve-year-old granddaughter Kaylee repeated a speech she gave in front of him when he visited her school a year before his death. 'I have a dream that one day, on the steps of Parliament House, the sons of former James Hardie workers and the sons of former James Hardie owners, will be able to sit down together at a table of brotherhood,' she fantasised.

Underneath Olympic Park and the stadium where Kaylee spoke lay tonnes of Hardie asbestos waste, dumped there in the late 1980s. The drivers contracted to spread the waste had been covered in dust. At least one later died from mesothelioma.[25] Yet on this sad occasion, John Reid, whose family had owned James Hardie for nearly a century, the man at the helm of the company when the waste was dumped and when Banton inhaled his fatal dose, was nowhere in sight.

BURSTING THE
PR BUBBLE

My interest in asbestos was first sparked in 1977 by a brief telephone call from a boutique public relations company, Neilson McCarthy & Partners. Would it be possible, the caller asked politely, for its client to use part of a recent radio program I had made 'for internal and external community relations purposes'?

The request intrigued me. 'Who is your client?' I inquired.

'James Hardie Asbestos,' came the answer.

I was working in the ABC Radio Science Unit, producing a weekly program broadcast on Radio National, an adventurous and progressive outlet with a relatively small audience. At the time I knew very little about asbestos. I had no idea that the broken fibro sheets I'd played with as a kid contained asbestos fibres which could end up in my lungs, and even if I had known, it would have made little impression. As a boy, I had seen photos of firemen in asbestos suits walking through flames, apparently unscathed, and I thought asbestos was some kind of miracle material that could protect against fire. That was about it.

But the call from the PR agency rang alarm bells. It was unusual in those days to receive a corporate request of this nature and I asked for confirmation in writing. I immediately began rummaging through the cupboard to find the original program tape, which I played back to discover what could be of such interest to James Hardie. I had no idea what I was looking for, other than it most likely mentioned the word 'asbestos'. I had to listen to the whole forty-five-minute program twice before discovering the magic word. I found it in one short sentence.

'The asbestos industry in Australia is no longer a problem,' Gersh Major had said.

Major was a physicist with the School of Public Health and Tropical Medicine. Situated within the grounds of Sydney University, the school was actually a part of the federal Department of Health specialising in occupational health, and had been receiving a research grant from James Hardie Asbestos, something I found out only later.[1] Major was the school's industrial hygienist responsible for assessing workplace safety.

In the twenty-minute interview I had recorded with Major I hadn't even noticed his comment. But Hardie's ever-vigilant spin machine had.

That really gave me pause for thought. 'No longer a problem.' Why would Hardie be so sensitive about its image that it would pounce on such a tiny, almost throwaway sentence by a public servant and seek permission to reproduce it? And why would it be paying a public relations agency to monitor the media so assiduously? In any case, why were these words useful for the company, unless there really was doubt about the industry's safety? Perhaps there still *were* problems. And who was this James Hardie Asbestos anyway?

During the following months I read everything about asbestos I could lay my hands on. I discovered that the industry in Australia had consisted of two major manufacturers, James Hardie and Wunderlich. Hardie was the larger firm, producing asbestos building products, insulation, and brake and clutch linings. While I was doing my research Hardie became an effective monopoly in Australia, buying Wunderlich from CSR, which had also operated an asbestos mine at Wittenoom in Western Australia. I learnt about the overseas industry, which had developed into a vertically integrated global cartel dominated by the UK's Turner & Newall and Cape Asbestos, Europe's Eternit, and Johns-Manville from the US.

But although I searched the newspaper files I found very few articles on asbestos, other than a few references to the overseas industry. I visited trade union offices, where I was allowed to look through their files and copy the ones I wanted, despite the initial suspicion the union officials had of a young, long-haired ABC journalist. One exception was the Sydney office of the giant Federated Miscellaneous Workers' Union. The 'Missos', as they were called, were initially offhand, but later downright unfriendly. This puzzled and frustrated me, because the Missos had by far the largest coverage of asbestos workers. I privately vowed to check out the union more thoroughly one day.

It didn't take long to discover that, contrary to Major's assertion, even in 1977 there were still huge problems in the Australian asbestos industry. Workers continued to be exposed to lethal doses of the dust. Companies were not complying with the law, in those places where there were laws. And there was widespread ignorance about the dangers of asbestos. People were dying from their exposure, uncompensated and in agony.

Not, however, according to many of the experts I interviewed. Gersh Major, lung specialists such as Professor

Bryan Gandevia, public health officials, and others I met through the NSW Dust Diseases Board all stressed how safe the industry had become. Some spoke about 'problems in the past' when the dangers of asbestos had 'not been recognised'. But those days were now over, they said. They used medical terminology seemingly as a weapon to intimidate rather than explain. Instead of talking about lung disease caused by dust, they said 'pneumoconiosis'; lung cancer was 'bronchogenic carcinoma'. I had to get my head around a baffling array of technical terms, and wondered how the sick factory workers dealing with them understood what they were talking about.

I learnt about the three main types of asbestos: blue asbestos was 'crocidolite', brown was 'amosite' and white, the most common, 'chrysotile'. Blue, which had been mined at Wittenoom, was the dangerous one, the experts told me, because it caused mesothelioma, a cancer that tragically only became apparent thirty or more years after exposure. Since blue had been effectively banned, they said, there should be no further exposures.

All types of asbestos could cause the disease 'asbestosis', which was essentially the scarring and shrinkage of lungs as they clogged with asbestos fibres. People with asbestosis breathe with difficulty and it is 'progressive', meaning it gets worse even after exposure to asbestos dust has stopped. The fibres keep digging in, causing more and more scar tissue in the lungs and reducing the room left for breathing. The heart has to pump harder to get the same amount of oxygen. As a result sometimes fingers and toes swell, or 'club', because of poor circulation.

I soon learnt to take lollipop steps when walking down the street with people who had asbestosis, because they had to stop every few minutes to catch their breath. Harry Dowdell, who worked at the Munmorah Power Station near Newcastle in NSW, had acute asbestosis. When he came to see me at the ABC with his workmate, union delegate Mick Smith, the stairs to my office were beyond him. He was puffed even walking from one room to the next. Yet he was only forty-nine. These two men and others like them told a different story from the 'experts'. They brought with them photocopies of overseas articles that talked of cancer. I began to realise that the subject would probably need more than a single radio program to explain. It felt more like a series. In fact, I spent a large part of the next eighteen months producing a series of four documentaries as well as a number of other shorter programs on the industry.[2]

By telephone I tracked down some of the Wittenoom workers, mostly recent immigrants, who were dying from mesothelioma. When the giant sugar company CSR began diversifying into building products in 1943, it had bought the Wittenoom asbestos mine in the hot desert country of Western Australia's Hamersley Range from mining magnate Lang Hancock.[3] Despite repeated warnings from state authorities about the dangerously dusty conditions, CSR only closed the mine in 1966 when the market for its fibre dried up. By then Wittenoom's deadly blue asbestos had already made its way across the country in asbestos cement sheets, insulation in power stations, factories and ships, and countless other products and uses, big and small.[4]

The more I delved into the industry, the more I realised how important it was to reveal what was really happening. I became conscious of soft, subtle lies that were being promoted about the safety of asbestos by a shadowy PR network which monitored media coverage worldwide. Later, I discovered that I wasn't being paranoid. The international PR company Hill & Knowlton was indeed coordinating an industry campaign. In Australia, James Hardie Asbestos was equally active. Later, I discovered in the company's archives that before I approached the company for an interview, its medical officer, Dr Terry McCullagh, was already monitoring my work. He had even written a confidential memo to the company's general manager, environment, tracking my progress:

> Mr Peacock of the ABC seems to have been most active in developing his proposed program on asbestos ... He has been in touch with Dr Longley of the Dust Diseases Board. Dr Longley tells me that Mr Peacock has studied his subject thoroughly and is very well informed. I gather he pressed on Dr Longley the view that chrysotile [white asbestos] and amosite [brown asbestos], if not equally potent causes of mesothelioma, were at least important causes of this condition ... Mr Peacock has also visited Perth where he discussed matters with Dr Heyworth ... Dr Heyworth was under the impression that Mr Peacock had visited ... Dr Selikoff in New York ...[5]

In fact, I hadn't visited Perth, only spoken by phone to people there. Nor had I yet visited Selikoff, though we had exchanged letters and I had interviewed him from the ABC's New York studio. Professor Irving Selikoff and his team at New York's Mount Sinai School of Medicine had enlisted the support of the US union covering asbestos insulation workers and had tracked a cohort of its members for thirty years. In addition to asbestosis, they had discovered a massive excess of lung cancer, mesothelioma and gastrointestinal cancer among the workers.

Their alarming results had dominated a 1964 session of the New York Academy of Sciences on the biological effects of asbestos, with the proceeds published and widely circulated the following year.[6] In Britain and the US, by the early 1970s the publicity given to his findings had galvanised asbestos activists, scientists and unions into a campaign to alert people of the hazards. In Australia there had been a resounding silence.

Selikoff warned Australian listeners for the first time that exposure to *any* type of asbestos — blue, brown or white — increased their cancer risk. He also criticised the official 'safe' level of exposure to asbestos. In Britain this level was set by regulation at two fibres per millilitre every eight-hour shift for a working life of forty years, based on a 1966 study using data supplied by Turner & Newall of workers at its Rochdale asbestos textile plant.[7] This study found that only 7 per cent of the workforce exposed to such a level would be at risk of developing asbestosis. But another study of the same workforce only a few years years later discovered that nearly half the workers had contracted asbestos-related disease.[8]

Selikoff had expressed his alarm to the UK industry about the standard: '… hundreds of thousands of men and women … are at serious risk of irreversible, often fatal disease,' he had warned.[9] He explained to me that the British standard was only supposed to prevent asbestosis; it made no pretence of protecting against cancer, for which there was no known safe level. In fact the standard of two fibres per millilitre still allowed the inhalation of an extraordinary amount of asbestos, because a single fibre of asbestos could splinter longitudinally into thousands of tiny shards, so small they could only be detected under a powerful electron microscope. We know now that the 'safe' standard meant breathing in about one billion fibres every working day.[10]

I found a series of articles about Selikoff written by Paul Brodeur, a journalist with the *New Yorker*, who had by then expanded his work into a book, *Expendable Americans*.[11] The book was a devastating exposé of the asbestos industry's efforts to conceal the scale of the emerging tragedy. Brodeur, in his interview with me, provided some alarming examples of the US industry's behaviour.

He had gained access to the files of Sumner Simpson, the former president of Raybestos-Manhattan, a major US asbestos firm, and he read to me some of what he described as 'the smoking gun' memos that he had unearthed. One had been written in 1949 by Johns-Manville's Dr Kenneth Smith.

A survey Smith conducted of 708 workers in the company's Canadian mine in the town of Asbestos had found only four with healthy lungs, but Smith nonetheless recommended the men not be told of their condition:

> As long as the man is not disabled it is felt he should not be told of his condition so that he can live and work in peace and the company can benefit by his many years of experience. Should the man be told of his condition today there is a very definite possibility that he would become mentally and physically ill, simply through the knowledge that he has asbestosis.[12]

Across the Atlantic, Nancy Tait had just published a book with the blunt title, *Asbestos Kills*. Her husband had died from mesothelioma, and I interviewed her about the UK industry's behaviour, which appeared to be equally outrageous.

Remarkably, almost none of these US and UK stories about asbestos had filtered through to the Australian media. One journalist, Timothy Hall, had written a prophetic cover story for *The Bulletin* magazine in 1974 headlined IS THIS KILLER IN YOUR HOUSE?,[13] but his was virtually a lone effort. It was only because ABC Radio National had pioneered quality international hook-ups around the world that I was able to reach out beyond Australia to gather interviews from overseas experts.

But as I discovered later from Hardie documents unearthed during various court cases and talking to company insiders, I was not the only person in Australia to be gathering the information. Inside Hardie's Asbestos House a group of its directors and senior staff had been carefully monitoring these developments and discussing them with their overseas colleagues. Selikoff's findings had been circulated to top executives, including Hardie director Frank Page, who found them 'reasonably interesting but not terribly informative'.[14] He had also read Brodeur's articles and book, which he described in a confidential memo as '... most depressing reading ... allegedly giving the victims' side of the evils of asbestos'.[15]

From Brodeur and Selikoff I learnt of the scandal that had just enveloped the Canadian industry, then the source of most of the world's white asbestos. Paul Formby, a young theology student and union organiser in the mining town of Thetford, Quebec, had grown tired of picking asbestos fibres out of his beard. He sought help from Selikoff's Mount Sinai staff in New York, who taught him how to take dust samples. Formby returned to work in the mines and secretly began to monitor the levels. When his results became public they triggered a national scandal, because they massively exceeded the recommended Canadian safety levels, which were later revealed to be inadequate. Formby's revelations provoked a bitter six-month strike by asbestos miners and a government inquiry.

Hardie sourced most of its white asbestos from Canada, although not all from Quebec. It also had a substantial share in the Clinton mine in British Columbia and was able to ride out the strike without disruption by drawing more on stocks there.[16] However, I discovered that Woodsreef, the major asbestos mine still operating in Australia, was Canadian-owned, too, and Hardie was its major domestic customer.

Chrysotile Corporation's Woodsreef mine near Tamworth, in northern NSW, had been put into receivership only one year after it commenced operations in 1972. For its continued survival the mine survived on a diet of government handouts. The NSW government provided a royalty holiday and the federal government offered substantial concessional loans.

When I visited the mine in 1977 to interview Harry Robinson, its smooth-talking Canadian manager, he declared with a straight face that the six-month strike in Canada was not about the asbestos 'health issue', but rather about money.[17]

As we walked through the Woodsreef mill I could see workers ahead of us hastily sweeping and shovelling dust, which lay inches thick on the floor. I tried to hold my breath! In our interview Robinson admitted there were 'some' safety problems at the mine. These problems would be fixed once the company could buy some new equipment, he told me, but agreed it was '... a matter of what you could afford'. He told me the link between asbestos and cancer had yet to be proven. Indeed, every

other week the specialists were coming up with the view that
'... this or that substance causes cancer', he said, with a twinkle
in his eye. I discovered years later that he had discussed the
Canadian government inquiry only a few months before and
had commented that the Canadian industry feared the
recommended 'safe' standard for asbestos exposure might be
lowered as a result of the scandal.[18] It was a conversation he
conveniently overlooked when I interviewed him.

I left my questions for James Hardie Asbestos until the end.
Hardie was the engine house of asbestos use and production in
the country and region, responsible for almost three-quarters of
Australia's asbestos consumption.[19] It was also the industry's
largest employer, with factories in all Australian capital cities
(other than Darwin and Hobart), in New Zealand's Auckland
and Christchurch, and joint ventures in Indonesia and Malaysia.

When I asked for an interview, Hardie chose to provide me
with two senior executives, Dr Terry McCullagh and Ray
Palfreyman. Palfreyman was head of James Hardie Environmental
Services, a title that suggested he was the person charged with
smoothing away concerns about asbestos. Palfreyman was also a
director of the company's asbestos-manufacturing subsidiary,
James Hardie & Coy, and would later die from mesothelioma.
McCullagh was a director of another Hardie subsidiary, the brake
company Hardie-Ferodo, and appeared intent on dazzling me
with his 'expert' knowledge of science and asbestos. They made
an intimidating pair across the studio desk.

Aware that I might not get another chance, I planned to
conduct a wide-ranging and tough interview to last about an hour.
At my side was a pile of research papers, including a confidential
company memo about the provision of respirators to employees

that I had discovered in one of the trade union files. Written that year by the company's manager of factory operations, Harry Hudson, the memo offered a damning glimpse into the company's strategy to protect its workforce. The Australian 'safe' standard for asbestos dust at the time allowed for exposure to twice as many fibres per millilitre as the UK, although Hardie publicly boasted that it used the tougher British standard of two fibres per millilitre in its factories. But as Hudson's memo indicated, when it came to the prospect of the company's workforce donning equipment that underscored the danger of the job, Hardie was ready to ignore its own standard by restricting the use of respirators. With the tape rolling, I read out the memo to the two executives sitting opposite:

> Heretofore it has been the practice to issue a respirator to any man who requests it. This practice is now contrary to policy. It is desirable that the new policy should be introduced as quickly and as completely as possible. However, discretion should be used. If a man requests the issue of a respirator despite the fact that his exposure is below four fibres per millilitre per shift, every effort should be made to persuade him that he has no need of the respirator, and it may be useful to seek the help of officers of the State Government Health Departments in this endeavour. On the other hand, the matter should not be allowed to develop into an industrial dispute, and if this seems a possibility, a respirator should be issued.[20]

McCullagh's response was haughtily dismissive. 'Well, let's face it,' he said, 'people in lofty positions sometimes tend to write memos that are not frightfully practical.'

Palfreyman was silent.

McCullagh clearly knew that both standards were unsafe, whether the dust count was measured at two or four fibres per millilitre. He had actually helped to set the higher Australian standard as a key member of the National Health and Medical Research Council's occupational carcinogens sub-committee. Another member of the five-person committee was Gersh Major, the industrial hygienist who had told me 'the Australian asbestos industry is no longer a problem'. The committee had doubled the UK standard for Australia on the dubious assumption that Australian employees would work for only twenty years, half the forty years their British colleagues were expected to work.

But both standards failed to protect against cancer. Indeed, more than a decade before my interview, McCullagh had told a conference of Hardie factory managers in July 1966 that: 'Recent literature has reported fairly conclusive evidence that asbestos dust when inhaled can cause lung cancer and cancer of the chest cavity as well as asbestosis ... There is no safe upper limit for asbestos dust.'[21]

Hardie's reaction was immediate when the 'Work as a Health Hazard: Asbestos' series first went to air on ABC Radio National's *Broadband* in July 1977. The public airing of the confidential memo led Hardie's Environmental Control Committee (on which both McCullagh and Palfreyman sat, as well as Hardie's in-house PR adviser Ron Bolton) to tighten security procedures for documents. In its minutes, the committee noted that Brisbane power station workers had been 'reassured' about the safety of their work with asbestos by Queensland Department of Health's Director of Industrial Medicine.[22]

Footage of Hardie's two factories in Brisbane had also

featured in a follow-up story to the radio series on ABC TV. David McPherson, a delegate for the manufacturing workers' union in one of the Brisbane factories, called me after the broadcast to say his manager had suggested that following this unwanted publicity they 'play it cool ... because all of our jobs depend on it'. McPherson was nervous about keeping his own job, but he was also scared about his health. The rafters in the factory were covered in dust, he told me, and when they fixed the ducting the dust 'goes everywhere'.

At the next meeting of Hardie's committee, concern was expressed that I might investigate Hardie's offshore work practices. 'Mr Bolton observed that Mr Peacock knew of our operations in Ipoh and Jakarta and suggested that it might be wise to hasten our education programme in those locations.'[23] At a later meeting, a Hardie executive declared that '... following the recent blaze of publicity there was now a very definite urgency in developing a dustless method of cutting Asbestos Cement.'[24]

In Sydney, the ABC's newly established young people's rock radio station, 2JJ, also followed up the Radio National series by editing my original programs into shorter segments and playing the grabs between music tracks.[25] Within days I received an angry phone call from the PR agency Neilson McCarthy: 'You didn't say it was being broadcast on 2JJ!' Hardie was obviously concerned that 2JJ had a different and larger audience to Radio National.

Little did the Hardie spinners know they had cause for greater alarm. The story was spreading through an even more effective medium. The Australian Metal Workers and Shipwrights Union (later called the Australian Manufacturing Workers' Union, or AMWU) had just established a sophisticated audiovisual library and its librarian had recorded my programs. He duplicated hundreds of cassette copies which were then distributed to AMWU shop stewards and delegates across the country. The programs were reaching the very people most likely to come directly into contact with asbestos. I received more worried phone calls from workers as far away as Western Australia who had heard the cassettes during their work breaks and wanted more information.

My interest in the Woodsreef asbestos mine was sparked again when I learnt that the Industry Assistance Commission (IAC) was planning an inquiry into whether the mine should continue to receive government assistance. Having seen the dust, inches deep, on Woodsreef's mill floor, I sought a response from the NSW Minister for Mines, Pat Hills, on how the government justified its subsidy when the conditions at the mine were so evidently unsafe.

Hills' media adviser arranged a meeting, saying, 'To be perfectly frank, Matt, the minister is not at all happy with them.' At the meeting in the minister's office we were joined by the Chief Inspector of Mines, Bob Marshall, who had brought along a very thick manila folder which turned out to be the department's file on Woodsreef. Marshall leafed through the file, observing that his inspectors had continually warned Woodsreef to improve its dust control. The mine's Canadian operators, however, had pleaded poverty, claiming marginal financial viability and had asked for a few more years' grace to upgrade their equipment.

After about fifteen minutes, the minister's media adviser looked at his watch and exclaimed that it was time for Hills to attend a luncheon appointment. I was welcome to stay and browse the file if I wanted. As the minister and his staff trooped out it suddenly dawned on me what was happening. As a young journalist, it was the first time government documents had been leaked to me. I poked my head nervously out of the room. There was an enormous photocopier next to the door. About forty minutes later, heart thumping, I bounded out of the building with a complete copy of the file, which I spent the next two days absorbing. I was astounded to discover that Woodsreef had never complied with the already dubious Australian asbestos standard.

I prepared a special radio report on the mine that went to air on the eve of the IAC hearing. It drew an angry response from Woodsreef's manager, Harry Robinson, whose PR agency wrote to the ABC Science Unit's director demanding a meeting to discuss my report's 'inaccuracies and distortions'.[26] In my absence, Robinson met with both my director and the controller of ABC Radio, but it was difficult for him to challenge a complete copy of the Mines department's file and I heard no more about it.

Initially the IAC recommended against subsidising Woodsreef, but its advice was overturned in the federal Cabinet following protests by senior National Party ministers Doug Anthony, Minister for Trade and Resources, and primary industries minister Ian Sinclair, in whose electorate the mine was located.[27] The NSW Labor government followed the federal lead and continued to provide subsidies and waive mining royalties. Woodsreef operated for another four years until declining demand led to its closure. Despite warnings sounded during the IAC inquiry, when the mine closed no provisions were made for follow-up health checks of its workforce or remediation of the site. Dust from its huge mountains of asbestos tailings still blows in the wind to this day.[28]

I discovered that the year before I began researching the radio series, Woodsreef's cash-strapped owners had purchased another, smaller asbestos mine from James Hardie. It was situated at Baryulgil, near the town of Grafton in northern NSW. Most of Baryulgil's workforce was Aboriginal. I can still remember my hair prickling on the back of my neck when I drove into the small settlement and noticed that the dirt road had turned white. Apprehensively, I held up some of the dust from the road to the sunlight for closer scrutiny. Sure enough, the road surface was asbestos dust!

In the school playground I noticed a large mound of asbestos that obviously served as the local equivalent of a sandpit for the children. My tape recorder picked up a background noise of dry, hollow coughs as the families there told me that they knew nothing about the dangers of asbestos. Dr McCullagh had taken X-rays of the miners every so often, they said, but he had never told them about the dangers, nor that there was anything wrong with their health.

On my return to Sydney I alerted other journalists before putting a radio report on Baryulgil to air, and within days the story was front-page news and featured in prime-time TV news bulletins. Hardie was now on the back foot trying to avoid the damning media images of Aboriginal children playing in the company's asbestos dust. It went to ground, refusing requests for comments or interviews.

Everywhere I looked, it seemed, there was another program to be made about asbestos. Melbourne's fleet of blue Harris electric commuter trains became my next story after I received a tip-off from a NSW Health Commission official, Eva Francis, that its carriages were lined with blue asbestos. When I suggested at the weekly editorial meeting that I follow up this tip, a colleague,

Malcolm Long, raised his eyebrows. 'Not another asbestos story,' he groaned. To jolly him along at the next meeting I hung a hunk of Wittenoom asbestos, sealed in a plastic bag, on a long string from the ceiling so that it dangled in front of his seat. As he sat down I said with a smile: 'If you think you've had enough asbestos, open it up and take a good sniff.'

In Melbourne I met a union organiser for the train drivers and at daybreak we secretly clambered over the Harris trains in Victoria's Jolimont rail yards. We took photographs of broken wall panels which exposed large tufts of raw blue asbestos only inches from where the passengers sat. Surely the passengers were at risk as they rattled to and from work in the cold Melbourne winter, windows closed and blue asbestos dust wafting through the carriages?

Victorian Labor Party shadow transport minister Tom Roper thought so. After the program aired he demanded that the trains be removed from service. But Melbourne transport would have ground to a halt if the government had heeded his advice, and within days he had changed his tune. The Harris carriages were inspected and 'sealed', and continued to operate even after Roper became transport minister in 1985. Three years later they were withdrawn from service, by which time the challenge was how to dispose of the trains safely. They were regarded as too dangerous to crush and eventually were buried in their entirety under sandpits at Clayton.

I soon grew adept at recognising asbestos insulation, which had been sprayed like a blanket on the ceilings of scores of public and private buildings as a fire retardant until the mid-1970s. Indeed, I had developed a habit of quickly glancing at the ceiling whenever I entered a building to see if it was there. I sent samples off for analysis and had soon outed a succession of Sydney's public buildings: the state library, the art gallery, the Australian Museum and Parliament House. Often the sprayed asbestos was deteriorating, with occupants complaining that flakes floated down from the ceiling as they worked. Subsequent monitoring of some of them revealed unsafe levels of dust. Closer to home, I interrupted a workman installing a false ceiling over similar sprayed insulation at the ABC's TV studios at Gore Hill. The ABC soon found itself trying to justify its actions to a group of angry current affairs journalists.

A teacher called me and told me that the ceilings at her school had also been sprayed with asbestos insulation. I discovered it had been common practice in schools across the country. When I reported the potential hazard on the ABC radio current affairs program *AM*, the story sparked a national reaction. Hardie's media monitoring began to creak under the strain. I found out later that there were so many stories about asbestos hazards the company's Environmental Control Committee required a separate attachment logging all the media coverage attached to the minutes of its meetings. The company hired Eric White, the local agent of the international PR firm Hill & Knowlton, to help deal with the deluge.[29]

Its consultant, Bill Frew, noted: '... the ABC's *AM* program broke a story on sprayed asbestos ceilings in schools which flowed over into the daily press and has led to an investigation of the ceilings by governments in three states ... So far in the media generally, James Hardie asbestos cement products have not been specifically identified as a potential health risk.'[30]

But it did not matter that Hardie was responsible for fibro cement sheets and not sprayed ceilings. Journalists did not make such a fine distinction and Hardie, the company whose headquarters still proudly bore the name Asbestos House, was an obvious target. Indeed, Hardie's media log had noted that a presenter on 2JJ, when about to play a track titled 'Living in Shame', announced 'We dedicate this song to those who work in James Hardie.'[31]

A crucial program in my original radio series had focused on the lack of cautionary labels on Hardie's products, an issue that galvanised protests from consumers, health bureaucrats and activists. Hardie had steadfastly refused to warn the builders or home renovators who purchased its fibro sheets at local hardware stores that its products had the potential to kill when they were cut. The company had fought to exclude asbestos products from the US and UK which already carried printed health warnings. Head of Environmental Services Ray Palfreyman cautioned overseas associates that '... the Australian industry would not welcome the importation of such labelled products into Australia ...'[32]

The Australian Consumers Association's magazine, *Choice*, weighed into the labelling argument late in 1977, calling for regulations warning do-it-yourself and domestic users of products like asbestos cement sheets of their potential cancer hazard.[33] *Choice* published a large colour photograph of the warning label used in the UK. Hardie responded by accusing the magazine of promoting an 'unjustifiable scare', although privately Hardie's Dr McCullagh informed his colleagues that the *Choice* article was 'generally sound'.[34]

When I interviewed the magazine's editor, Geri Ettinger, she was unapologetic.

'We don't think it's unjustifiable at all. The consumer is very unaware of the problems and definitely a labelling scheme should come in,' she said.

The labelling controversy did not go away. Members from the Workers' Health Action Groups (WHAGs) in Melbourne and Sydney printed bright yellow stickers warning, DANGER: ASBESTOS! DUST FROM THIS PRODUCT CAN CAUSE CANCER WHEN INHALED, and threatened to fix the stickers on asbestos cement sheets as they came off the Hardie production line.

At Lidcombe, in Sydney's industrial belt, Workers' Health Centre members organised a rally outside Hardie's annual general meeting. About twenty demonstrators paraded outside Asbestos House, which they had renamed 'Asbestosis House' with the help of a spray can, and gave out leaflets calling on Hardie to place warning labels on its products. They also challenged chairman John Reid to reveal how many of his employees had contracted asbestos-related diseases.

Hardie PR staff learnt of the planned demonstration the day before the AGM and had briefed the chairman, John Reid, on how to respond to difficult questions he might be asked at the meeting. With the images of the Baryulgil mine still vivid, the minders agreed that 'some statement should be available for the Chairman relating to our earlier refusal to answer questions about Baryulgil'.[35]

At the meeting, Reid was asked about a payment to a Baryulgil widow, Ruby Mundine, who had received $3000 in compensation when her husband Cyril died.

'Not our problem,' said Reid. 'He was an employee of Asbestos Mines Pty Ltd, which we no longer own.'

Reid closed the meeting by thanking shareholders for bearing with '... the nonsense you have been bombarded with'.[36]

Following the meeting, at a press conference for the financial media, several journalists asked the chairman to quantify the number of Hardie's asbestos victims. Initially, Reid said he didn't know. When pushed, he claimed that 'in the past fifteen years one hundred of the company employees have contracted asbestosis and other related illnesses'.[37]

The Hardie spinners had also anticipated that the warning label controversy would be raised at the AGM. Behind the scenes, the company had begun to take action to defuse the mounting pressure. After concluding that the British industry had 'not lost any business on this issue', it had designed a label that carried the mild caution: '... asbestos dust can damage health ... but this is unlikely when cutting or fixing asbestos cement sheets since only a *small percentage* of dust is asbestos. As a precaution, however, you should keep dust levels down ...'

But the wording for the label was not considered strong enough by the NSW Health Commission, which insisted that 'cancer' be included, a suggestion Hardie vigorously resisted. Eric White recommended that Hardie ensure the labelling was voluntary and that it include the word 'seriously' as a concession to the Health Commission.[38] Reid, however, was not ready to concede anything, drawing a comparison with the tobacco industry, which was not required to mention cancer in its warnings. At the AGM he stressed there was '... no danger to the home handyman cutting asbestos cement, provided the simple steps printed on the [Hardie] sales brochure are taken'.

Within days, Eric White repeated its advice, urging Hardie to press ahead quickly with its own labels:

> In taking the initiative, the company can quite rightly point out that delays to date have been the result of protracted debates with the bureaucracy over the wording to be used … It [Hardie] could say, 'We think it's high time that the consumer was given some advice on the safe use of the product and that the label we are using is better than none at all. If officialdom wants to debate the wording with us later on, we are happy to do so. However, in the meantime, the consumer will have been afforded some protection.'[39]

The company relented, heeding the advice of its PR consultants, and pressed ahead with its own labels. Through its advertising agency Coudrey Dailey it commissioned market research to test the reaction of its Sydney and Melbourne customers. The results revealed no adverse reaction to what Hardie privately conceded were 'innocuous' labels.[40] Hardie's in-house PR manager Ron Bolton authorised further research in 1979, with the agency noting: 'In the first study no mention was made of the word 'cancer'. It was felt this now should be included.'[41]

Mild though the Hardie warning was, the company's new-found concern did not extend to all its customers who might be at risk. Hardie soon followed the lead of Turner & Newall in the UK by placing labels inside its Hardie-Ferodo brake packs, but '… railway brake blocks are at this stage not going to be labelled, because of the sensitivity of the unions involved'.[42] And in Indonesia and Malaysia, Hardie's products still had no labels in 1980, when an industry spokesman said he was sure they would be labelled 'in due course' but that '... labelling ranked lower in priority than dust extraction and worker check-ups'.[43]

The NSW Health Commission continued to demand tougher wording on the labels, with Eric White recommending meetings with individual state health ministers to counter the 'extreme positions' held by the NSW officials.[44] Hardie was able to stall a cancer warning for another four years before it was recommended by the national health authority. The company reluctantly complied, still complaining it had been unfairly singled out. By 1982, though, the issue was becoming irrelevant because the company was already partway through eliminating asbestos from its building products altogether.

The process of abandoning asbestos had already begun, in name at least, following the demonstration outside Asbestos House in 1978. Within a year, James Hardie Asbestos changed its name to exclude 'the contentious word 'asbestos''.[45] At the same time the company, now James Hardie Industries Pty Ltd, took down the brass plates at its headquarters in Sydney's York Street. 'Asbestos House' had been erased.

I was about to leave the ABC in Sydney — and the asbestos story — when I received a call from a former Hardie engineer, Fred Sandilands, who had worked for Hardie most of his life. His most recent job was as an engineer on the dust committee which oversaw clearing hazardous asbestos dust from the factories. Aged forty-nine, he had just learnt he had mesothelioma.

Sandilands was intensely proud of his time with the company and spoke of his affection for Thyne Reid, the engineer who had chaired James Hardie Asbestos until 1964. Reid had been an eccentric character who would land his gyrocopter at the factory and, wearing a necktie as his belt, was often seen on the factory floor fiddling with equipment and dreaming up new designs.

Sandilands talked about the former practice at Hardie's Camellia plant of disposing of the 'fines' — very fine asbestos waste — by dumping them throughout the Parramatta area. He was prepared to go public on television with his concerns about the safety of the dumps, so I contacted Mark Colvin, then working for the ABC TV current affairs program *Nationwide*. Colvin filmed Sandilands pulling out asbestos waste from embankments and other dump sites. Sandilands remembered the spots well because he had sometimes supervised these operations. He estimated that in the 1960s alone Hardie had dumped at least 12,000 tonnes of asbestos dust and sludge in the area and at least 120,000 tonnes of broken asbestos cement pieces. On occasion, Hardie workers had covered the waste with soil, but Sandilands was easily able to demonstrate for viewers that it had often worked its way to the surface and could be scooped up by hand.[46]

Hardie's response to this sensational story was derisory. The company doctor, Terry McCullagh, scoffed at the suggestion that the waste dumps could affect people's health.

'The chances of being hit by a bus are far greater,' he said.[47]

three

THE INSIDERS

James Hardie was good at keeping secrets.

When I first began to research the company, it was near impossible to penetrate its PR defences and find insiders bold enough to talk to me openly and candidly, especially managers high enough in the hierarchy to understand Hardie's asbestos strategies and corporate culture. Those who did call me with information after my early ABC radio programs were usually unwilling to give their names.

A notable exception was Peter Russell, who first contacted me in 1977 from semi-retirement on the Queensland coast. He was keen to talk on the record and promised to send me a bundle of Hardie documents, which was to prove so valuable that it later became a resource for lawyers suing Hardie on behalf of clients suffering from asbestos diseases.

Russell began working as an assistant in Hardie's laboratory at Camellia in 1948, while he was a chemical engineering student at Sydney University. The role of the lab team was to research and test Hardie asbestos products, such as its flat sheets for buildings, corrugated sheets for roofing, pipes for pumping water and sewage, and gas and water meter covers. He recalled that in those days it was impossible to avoid asbestos dust at the

Camellia complex. The milling area, where the raw asbestos was tipped from bags into the fibro mix, was particularly bad.

He told me that his colleagues at Hardie didn't regard asbestos as being particularly dangerous. Injuries at work were viewed as regrettable, but a fact of life. Workers even joked about their injuries and the compensation they were worth. Russell told of one young Maltese worker who had some finger joints missing. He would hold up his stumps, laughing: 'That was my motorbike, that was the deposit on my house ...'.[1]

The asbestos dust was viewed in the same jocular way. A 1940s cartoon in the company's in-house magazine showed workers from Hardie's Adelaide factory covered in asbestos fibre from an overflowing bin. It was captioned: *Birkenhead, the only place in South Australia where it snows every day.*[2]

In 1959 Russell was promoted to superintendent of the Hardie insulation factory at Camellia where a young Bernie Banton would work a decade later. Some workers under his supervision had already been 'dusted', as the condition of asbestosis was called. He offered a clue to the Hardie culture by telling the story of Percy Leabon, who had worked at the insulation factory for about thirteen years. After retiring sick, Leabon lodged a compensation claim for asbestosis that was unsuccessfully contested by the company's insurer, Manufacturers' Mutual. A Hardie document that Russell gave me read: 'He died in November 1959, aged 69. His wife made a claim for a lump sum settlement alleging that asbestosis was a contributory cause of death but she also died before the hearing and the matter lapsed.'[3] He remembered that the Hardie personnel officer, on learning of her death, placed the Leabon file at the back of a cabinet, saying brusquely, 'We don't have to worry about that anymore.'

Russell was appointed safety engineer and fire officer at Camellia in 1961. In his new job he focused on the hazards of asbestos dust with growing alarm. He soon learnt how to take dust measurements in the factories that made up the Camellia complex. He explained that the use of power saws was common in those days and some of the readings he took were extraordinarily high — thirty or more times the recommended standard of the day, which was itself inadequate, as he pointed out. He later told a court hearing that during this time he had felt like an 'accomplice to murder'.[4]

The documents that Russell mailed to me were an eye-opener. He had first seen them in what he called Hardie's 'Dust File', a manila folder that was handed over to him when he moved into the safety position.[5] On reading its contents, Russell realised that Hardie had secretly monitored the health of its workforce for decades, methodically keeping lists of the workers who had fallen ill from its asbestos.

One of these documents, 'The position regarding Asbestosis at Camellia, to the 1st April 1964', contained the names of twenty-three workers, listing their age, years of exposure, how and when their condition was discovered, the action taken and their current situation.[6] My attention was caught by one entry: Filipovic, aged fifty, had worked in the insulation factory for six years when a 1957 NSW Department of Health report recorded his condition as 'possible early asbestosis'. He was transferred to the asbestos cement factory as a cleaner. Four years later he was dead. The Hardie list noted: 'Death believed to be from "heart trouble following bronchitis".'

The inverted commas had caught my eye. Why were they needed? Was Filipovic ever made aware of his condition before he died? It was known that asbestosis could induce heart disease from the strain of pumping oxygen through lungs scarred and constricted by asbestos fibres. The lungs, in turn, could develop secondary infections like bronchitis. Three more men on the list were described as having 'heart trouble' — the kind with inverted commas.

The document revealed that in 1964 nine out of ten men employed in the so-called asbestos gang were identified in health department surveys as having some form of asbestosis. Members of the gang had the most dangerous job at Camellia: they unloaded bags of raw asbestos coming into the factory, poured the fibre into bins, and mixed in different proportions of white, brown or blue. The tenth worker, Ernie Schofield, had presented a local doctor's certificate stating he had a heart condition caused by lung disease. When Russell had observed to a colleague that Schofield obviously had asbestosis, he remembered being abruptly rebuked: 'Who are you to question his doctor?' After further inquiries, he located Schofield hidden away in an obscure storeroom where, he told me, he presented a 'pitiful sight ... sitting embarrassedly behind the shelves and shuffling out in his slippers when someone came in'.

Russell suggested to management that Hardie should encourage autopsies for workers who died to establish their cause of death beyond doubt, but he got nowhere. He began to think about it more. How many people really were getting sick? The large migrant workforce at Camellia had a high turnover and signs of disease were everywhere among those who stayed.

On one occasion he wrote a list off the top of his head of the (mainly supervisory) staff who could have developed an asbestos disease. Thirty-six names are written in ink on the document and Russell wrote a note for me explaining that the list was by no means complete. Eleven names were marked 'A' for asbestosis; six names had a 'C' for cancer; and ten names had an asterisk signifying heart conditions. He had underlined the twelve names of those who had already died. Drawing up this secret tally of affected colleagues must have been a devastating experience for him.[7]

Russell began to do his own research about asbestos at the Mitchell Library in Sydney, where he was to discover a series of overseas studies on asbestos. Among the studies was the 1930 landmark research by Merewether and Price in Britain which resulted in regulations for the UK industry designed to reduce future unsafe exposure.[8]

What Russell didn't know was that as a result of the British regulations authorities in Australia launched their own investigations. At Hardie's Riverdale factory in Western Australia, the Chief Inspector of Factories identified two Hardie workers — one employed for six years and another for ten — who in 1935 appeared to be suffering 'in a marked degree from the effects of asbestos dust'.[9] Within a year of this inspection, James Hardie Asbestos moved all its asbestos business into a subsidiary company, enabling the parent company to claim the legal protection of the 'corporate veil' against future asbestos damages suits.[10]

Nor did Russell realise that four years later, in 1939, the company was first sued for death caused by asbestos dust.

Samuel Jones had been employed at Hardie's main asbestos cement factory at Camellia for seventeen years when he started to have trouble breathing. His job was to rake out the raw asbestos, which made the dust billow up and clog his mouth and throat. Racked by a bad cough, he lost his appetite, rapidly lost weight, and died soon afterwards. His widow's claim failed, despite evidence from a doctor that the dust conditions in the factory indicated that an asbestos hazard was possible. It was assumed that he had developed his condition from his earlier occupational exposure in Welsh coalmines. Evidently unaware of the Merewether and Price study, the judge concluded that '... while, generally speaking, exposure to asbestos dust is an industrial hazard, there is nothing known as to what degree of exposure is necessary to cause asbestosis'.[11] The UK study had, in fact, set out to establish precisely what degree of exposure to asbestos was required to cause disease.

When I told Russell years later that Hardie's first asbestos compensation claim had occurred in 1939, he was quiet for a moment.

Did it surprise him, I asked, to find that Hardie had been sued so long ago?

He gave a dry chuckle. 'Well, it's par for the course, really,' he replied. 'The question you ask is, 'How did they get away with it for so long?' It just got covered up. It was all kept secret. James Hardie was fairly influential.'

As he continued to read documents in the public library, Russell learnt that even the asbestos dust standard set by the British study had proven to be woefully inadequate, with a follow-up study in 1955 demonstrating an excessive rate of lung cancer among the same workers.[12] He also noted that the research literature 'was starting to mention mesothelioma', then an extremely rare tumour. The disease was identified in 1957 among South African miners and their families who had developed it many years after only slight exposure to blue asbestos.[13] A few years later mesothelioma was also found in a miner from CSR's Wittenoom mine, where Hardie sourced thousands of tonnes of blue asbestos.[14]

Following his research, Russell wrote a memo to Hardie factory managers and senior staff warning that '... overseas, asbestos is classified as one of the most dangerous of industrial poisons'.[15] But he was a lone voice among the managers and felt his efforts were futile. In fact, the dust problem at Camellia appeared to be worsening as the factory expanded production.

He recalled an occasion when a colleague had screened an industry promotional film to foremen, supervisors and technical staff showing Wittenoom miners operating underground in very dusty conditions.

'I was disgusted,' he later commented. 'The showing of the film completely cut the ground from under me in relation to my efforts to get people in the factory to wear masks.'[16]

Russell had begun to confront the Hardie culture. He became the manager who spoke about a subject that others preferred not to discuss. As he explained to me, it was regarded as 'disloyal to openly raise these matters'. Yet he knew that he had some secret sympathisers because they would anonymously leave press cuttings about asbestos health effects on his desk.

At the library, Russell had found a promotional pie chart from the sales literature of the US asbestos manufacturer Johns-Manville that listed the extraordinary range of asbestos-based consumer products, from beer filters to hair dryers. 'If asbestos were really dangerous, then it struck me that all these people might be at risk,' he worried out loud. 'What if Hardie's customers were in danger?' He signalled his intention to raise the question at the next monthly management meeting.

To his dismay, the meeting was not conducted in the usual way: the chairman cut short its proceedings by announcing that Hardie would soon introduce a wet treatment process that would remove the dust problem. Russell said that the others present were already on their feet and heading for the door as he called out, 'Hold on, hold on,' and voiced his concern about the risk to Hardie's customers. Thumping the table, he urged that the company put a written warning on its products. 'We have a moral obligation to let the consumer know,' he insisted.

He was soon put in his place. Once the product was sold, he was told, Hardie had no responsibility for it. The branch manager took him aside after the meeting and accused him of causing embarrassment, saying 'I don't know what I can do with you, Peter.' He was to later die from mesothelioma, Russell added.

In fact, Russell wasn't the only person at the time to raise questions about the safety of the so-called asbestos 'end-users'. Unbeknownst to him a large Australian buyer of asbestos cement had already expressed its concern about the possible dangers of the product, a worry it conveyed to both the NSW Department of Health and the industry. In May 1954 the medical officer for the Snowy Mountains Hydro-electric Authority (SMA) had sounded the alarm that employees in the SMA's Cooma workshop were exposed to 'highly dangerous asbestos dust' as they cut fibro sheets for prefabricated housing.[17] In response to what it described as this 'serious health hazard', SMA management suggested that inquiries be made of Hardie, its fellow asbestos manufacturer Wunderlich and the NSW Department of Health. When the department's industrial hygienist, Harry Whaite, travelled to Cooma to investigate, he had found quantities of airborne asbestos dust in the workshop that were almost twenty times the recommended safety level.

Following Russell's stand-off at the Hardie management meeting, he attempted to raise the issue directly with more senior executives. He buttonholed other managers to discuss the health hazards, but was fobbed off. He approached Dr Robert Hughes from Hardie's insurers, Manufacturers' Mutual, who did not seem overly concerned. And Hardie's part-time medical officer, Dr Graham Kroll, was 'totally disinterested'. According to Russell, Kroll used to sweep into the factory once a week in his Rolls-Royce and had a handy sideline selling franchised protective creams to the workers. Russell even attempted to call the then chairman, Jock Reid. He left a message with Reid's secretary, but when he phoned back he was told Reid had been 'called away urgently'.

His next step was to take his concerns to the media. The Fairfax media empire was located next door to the technical college where he had transferred his studies. One evening in 1968 he walked into the building, ready to reveal his story, but he soon abandoned the attempt in despair.

'There was a little nook where there was a chap with a visor on his head. I said, 'I need a journalist to help me.' He told me to put the story down in 800 words and they'd look at it. Afterwards, I thought about the advertising account James Hardie would have with them, and I left it.'

On another occasion Russell spoke to a man in his fifties who was waiting outside the Hardie employment office. 'I asked him if he was aware of a few industries next door and further down the road. He said he had not much chance there as they snapped up all the younger labour and our place was known as the 'geriatric factory', which was why he was here.'

Russell knew that because of the long delay between exposure to asbestos and the onset of disease, Hardie's Dr McCullagh had actually recommended the employment of older men who would have 'less asbestosis, less lung cancer and less mesothelioma', and would be more likely to die from other causes because of the long latency period of asbestos diseases.[18]

Eventually Russell resolved to resign from Hardie. He penned another memo to senior staff detailing the hazards of asbestos, predicting that Hardie could become 'sitting ducks' for claims from sufferers of asbestosis, cancer and related heart conditions.[19] Only thirty-seven years old, he took long service leave while he considered his future. He was on leave when a friend working at Hardie's brake factory offered him a different job, which he accepted, but in 1970 he left the company altogether, citing in his letter of resignation his concern over the industry's dust problems.

Around the time of Russell's departure, Ron Hinton, a colleague of Russell's from his days working in Hardie's research lab, set off overseas with Hardie's chief chemist to swap information with Hardie's partners in the global asbestos cartel. They visited the huge factories run by Cape Asbestos in Britain, as well as the Belgium asbestos manufacturer Eternit.

'They were all doing experiments to try to find an alternative to asbestos,' Hinton told me. 'They knew that they ultimately had to get out of it.'[20]

Hinton had begun work with Hardie as a trainee chemist in 1945, and remained working in the lab until he was promoted to manager of Hardie's insulation factory in Sydney. In 1971, he moved to head office as an assistant to the director of production for Hardie's Australian asbestos cement operation. After helping establish Hardie's factory in Indonesia he became the operations manager for all of Hardie's overseas plants, which extended from Indonesia to Malaysia and New Zealand.

Unlike Russell, Hinton told me that neither he nor his colleagues ever spoke much about the actual dangers of being exposed to asbestos. Warwick Lane, who took Hinton's job as manager of the insulation factory, described it as 'a bit like getting killed on a motorbike. It was not the sort of conversation you would have over a beer. Sure, a few guys go down. It was a bit dangerous, the attitude was, so let's try to reduce the dust.'[21]

Looking back, Hinton knew people who fell ill — probably from asbestos-related illnesses.

I don't think they were even diagnosed. I mean, they probably died from lung cancer. But I can remember a bloke in the sixties in one part of the plant that I was in, the chemist, and he died from stomach cancer, and the way his stomach blew up, in modern thinking he would have had a mesothelioma. But you know, it was never related to the asbestos.

In Hardie's research labs tests were constantly carried out on different combinations of fibres and other ingredients added to the mix to improve the company's product. The company's chemists found that brown asbestos, because of its length and coarseness, greatly assisted in filtering the slurry of cement and fibre which was fed into the factory production lines, and brown was used in varying percentages for almost all of Hardie's asbestos cement.

Hinton said Hardie 'led the world' with some of its experimental work. 'We had intricate tests for each grade of asbestos and would vary the combinations to get different results.' In later years, when the company decided to get out of asbestos altogether, dozens of chemists, lab assistants and engineers worked with miniature machines to simulate the Hardie production lines.

During his time setting up Hardie's Indonesian business, Hinton had a lot to do with John Reid, who used to come to his office when he visited the country.

'He was an unctuous bugger,' Hinton said. 'He was born with a silver spoon in his mouth and he let you know.'

The Jakarta factory that Hinton supervised continued to produce asbestos cement after Hardie sold its share to its local partner, the Bakrie family, in 1987; after the 2004 Asian tsunami, people in Aceh protested when they discovered the materials provided by the factory to rebuild their homes still contained asbestos.

Hinton remained philosophical about the risk. Despite the loss of several close friends to asbetos diseases, he continued to work for the asbestos industry as a consultant into his seventies.

That was not the case with Neil Gilbert, another insider who was prepared to talk to me frankly many years after leaving the company. He had also begun work in the research laboratory and was to play a key role in developing the company's response to the health hazard.

Hardie sent him to Melbourne to install new machinery at its Brooklyn factory, where he was soon promoted to assistant works manager. As a young engineer, Gilbert had an interest in dust disease prompted by his experience as a boy in the West Australian goldmining town of Kalgoorlie, where his father had contracted silicosis from his work in the mines.

'I knew that dust could hurt you. Asbestos was a silicate and I inquired about it, but was told, 'Oh yes, we get asbestosis, but you take people away from it and they get better. It's no problem," he explained when I first met him.[22]

While in Victoria, Gilbert was handpicked in 1954 by Hardie director and chief engineer Frank Page to clean up the asbestos dust in its local operations following pressure from the health department. As Gilbert noted, his task was to ensure that the Brooklyn factory be 'as dust-proof as possible ... because the Department of Health is insistent that the operators at present are in constant contact with asbestos and a means must be sought to alter this'.[23]

Like Russell, Gilbert described how dust was 'everywhere' throughout the plant.

'We had about thirty-six men hand-mixing on the floor and tipping the asbestos down the hole into the treatment plant ... and the blokes on the blow room pulling the stuff out by hand ...'

Dr Bryan Gandevia from Melbourne University had already tested and identified signs of asbestosis in the heavily exposed workers and had optimistically (but erroneously) advised that their condition would improve if they were moved away from the dust.[24] Gilbert's solution was to install new equipment which automated almost the entire process, with only one man left to handle the asbestos directly, and the dust around him was drawn away by negative air pressure.

Aware of the growing fears about mesothelioma, Page subsequently decided to explore another solution: the complete replacement of asbestos. He instructed Gilbert, who by 1963 was the Brooklyn manager, to develop a cellulose alternative, an option Hardie had considered during the postwar years when asbestos supplies had nearly dried up.[25]

Page sent Gilbert a sample of a fibro board from Belgium. Called Menuiserite, it was made with cellulose and asbestos.

'I fiddled around with it and tried altering the constituents. I decided to make it a lime silica board ... including silica and cement to get a lime silica reaction when it was autoclaved [steam cured],' Gilbert told me.

After much trial and error, he came up with a combination that seemed to work. He had invented Hardiflex.

The official Hardie history later hailed this invention as 'a major technical breakthrough ... the first major step in producing an asbestos free fibro-cement'.[26]

Hardiflex was launched on the Victorian market, where, because of its increased flexibility, it was promoted as an alternative to asbestos cement.

Gilbert said that he was sure that Hardiflex could be made with no asbestos content at all. In fact, he had done just that. 'I had done some runs without any asbestos. I did get some delamination problems, but I didn't do any further work. I told the production manager, "I can get that asbestos lower." He said, "No, don't reduce it. We have to buy grade-three of asbestos for the pipes and we have to take a proportion of grade-five."'

Warwick Lane, who worked in Melbourne at the time, spoke to Gilbert about his experiment and recalled his account of this conversation.[27] He thought little of it. Hardie's pipe manufacture, although of a smaller scale than its asbestos cement sheets, had delivered half the company's profit since the early 1960s, and it was not surprising that asbestos was needed for the mix. Cellulose increased the flexibility of the product, but the fine-grade asbestos gave pipes their required strength. Hardie was able to purchase the fine-grade fibre at a discount if it also bought volumes of the lower-quality grade, which were then added to the mix for asbestos cement sheets.

Others at Hardie, like Hinton, although amused by Gilbert's experiment away from the stern eye of executives in Sydney, dismissed the idea that his Hardiflex sheets would have been strong enough without any asbestos. It was only much later, Hinton explained, that the company's chemists discovered the type of cellulose that, when treated, would provide the required strength. In those days, however, eliminating asbestos was not the company's research priority. When Hardiflex was launched nationally in 1964, it contained 15 per cent asbestos content and Hardie accelerated its campaign to market its pipes.

Gilbert was aware of the accumulating evidence that asbestos caused cancer, a fact he told me was 'well known' within the company by 1961.

'People were getting sicker. You could see it. It was a gradual thing, [but] you started to think: 'Bugger me, they don't seem to be getting better!' And then this cancer scare came up and people said, 'It's absolutely nothing to do with asbestos, it's just the smoking.' But the alarm bells were ringing. By '65 the company was fully aware that [asbestos] could cause lung cancer, particularly as a co-carcinogen with smoking.'

Gilbert moved back to Sydney in 1965 with new instructions to clean up the Camellia factory. Hardie was to spend more than half its capital budget at the plant that year on dust-extraction equipment.[28] The following year he was sent to the US to learn what steps Johns-Manville was taking to control asbestos health risks.[29] He was given full access to senior staff and factories, and was impressed by what he saw, although he felt the Johns-Manville solution was only implemented half-heartedly.

'It was more a PR cover ... a good scheme that hadn't been made to operate. They paid lip service to dust.'[30]

He returned to Australia with plans to install dust-extraction and monitoring equipment in every Hardie factory and also to establish a program to monitor the health of employees. He warned Hardie's board that the company would be out of business within ten years if it didn't establish the scheme and introduce an asbestos-free range of products.[31] The board accepted half his advice, but it ignored his recommendation to phase out its asbestos products for almost another twenty years because the profits to be realised from its pipe manufacture, which required fine-grade asbestos fibre, were simply too alluring.

Gilbert teamed up with the company doctor, Terry McCullagh, who now assumed a full-time role surveying the health of Hardie's workforce.[32] The doctor's estimate from his initial health screening was that two hundred employees had already been adversely affected by asbestos.[33] As he equipped the factories with dust-extraction machinery and dust monitors, Gilbert soon found himself involved in a battle with local managers, who resisted the push to lower the extraordinarily high dust levels at some factories. He and McCullagh gave what Gilbert described as a 'song-and-dance' routine in each city, explaining the dangers of asbestos and the purpose of the new scheme.

Gilbert claimed that he didn't 'mince words' at his factory

briefings, but when he arranged for the transfer of sick men from the dusty Camellia insulation factory, a memo he wrote cautioned management: 'Any transfer of men necessary should be carried out over a period so as not to excite undue comment.'[34]

Hardie's PR similarly avoided alerting people to the dangers: its public utterances still very much minimised the hazard. Again and again in statements prepared for the media or staff, Hardie would play down the dangers of cancer and suggest that any disease was the result of past exposures, when dust levels were 'much higher'.

At this stage, even Gilbert wasn't fully convinced of the gravity of the situation. He fought for lower dust levels, but found it difficult to believe that brief exposure to asbestos could cause mesothelioma. Like Russell, though, the more he studied the evidence the more apprehensive he became.

> The crunch started to hit in 1965 and 1966. Could it be true one contact could do it? I thought, 'No, it couldn't be!' But gradually it had to be accepted, and suddenly have your mind accept that one sniff could do it! When you see thousands of people working with it and nothing happens to them, it took a big leap of faith to say, 'Well, that's it!' That took a lot of accepting, a helluva lot of accepting. You know, I went from 'bloody nonsense' to 'hope it's not true' to 'well, it must be!'

By 1966 Dr McCullagh had told Hardie factory managers that '... recent literature has reported fairly conclusively that asbestos dust when inhaled can cause lung cancer and cancer of the chest cavity lining as well as asbestosis ... there is no safe upper limit of exposure ... any exposure is dangerous and cumulative.' Nor was the risk restricted to asbestos workers: there appeared to be a danger for people '... living within half a mile' of an asbestos factory.[35]

Ominously, McCullagh warned, '... as dust levels reduce more cancer will become apparent ... with better dust control those being exposed are now living long enough to develop cancer.'

Blue asbestos had become the prime suspect for causing mesothelioma, but Hardie was aware that the other forms of asbestos might also cause the disease. When I later read back to Gilbert a letter he'd written to a former Hardie manager in New Zealand, in which he attempted to verify whether a worker with mesothelioma had ever been exposed to blue asbestos,[36] he conceded it had been a real worry.

'We were concerned that if it wasn't just crocidolite [blue], then we still had the problem with amosite [brown] and chrysotile [white], and we were trying to make sure that wasn't the case.'

Gilbert realised that even if only limited exposure to blue asbestos was the cause of mesothelioma, there were still going to be a lot of deaths. But blue, to the exclusion of the white and brown, would become a convenient culprit for the Hardie board as it began to shift ground over the cancer hazard. Hardie had agreed in 1957 to purchase CSR's blue from Wittenoom in order to forestall a tariff on its imported brown and white, and was now happy to abandon it. As director Frank Page wrote in 1968: '... on purely economic grounds, quite apart from any biological effect, we have no interest in WA crocidolite.'[37] He suggested that the company 'dribble' the remaining stock of blue through the pipe plant during the next few months to get rid of it.

It took another five years for the board to formally resolve that Hardie would no longer use blue asbestos. Hardie gambled that the highly toxic blue dust was the main cause of mesothelioma. But its brown dust was also extremely dangerous. Although scientists still argue about the relative mesothelioma risk from different types of asbestos, the evidence is now overwhelming that *all* forms of asbestos can cause mesothelioma as well as lung cancer.

Hardie's asbestos sales were booming, but Gilbert had come to the conclusion by 1969 that the company goal of 'engineering the dust out of existence' was unachievable.

'I was starting to realise we couldn't get dust down to the level required, which was nil. We had to go to products which didn't have asbestos in them.'

He urged senior Hardie executives to further reduce the asbestos content in Hardiflex, and had an abrasive confrontation with director Jonah Adamson on the subject when he insisted that the company needed to move faster.

A short time later he was no longer invited to meetings of the executive committee, and the job of running the medical and dust-mitigation scheme was handed over to the personnel manager, Ray Palfreyman. Gilbert resigned in 1971.

He shook his head over the company's refusal to phase out asbestos earlier. 'They hadn't done one thing I thought they should have done, which was to get the asbestos out of the product altogether, which they could do. They didn't see beyond the fact that [if they did phase out asbestos] they couldn't keep producing pipes.'

Peter Russell had reached a similar conclusion. He wrote to me in 1977:

The capital, the number of people involved and the extent of the industry were so large that there was an understandable reluctance for the owners, directors and senior executives to openly engage in debate on this controversial and agonising question of the future of the industry ... I cannot accept that this cursed material is irreplaceable and cannot be banned, at least in a dry condition ... A decision is unavoidable, and we must lock horns with it. Such a decision cannot be left in the hands of the industry.

four

THE 'DUBIOUS STATISTICS OF DEATH'

'It helps no one to go on accumulating dubious statistics of death,' wrote Hardie's Dr McCullagh in 1968, answering an inquiry from a state public health official about the fate of Hardie's former employees.[1]

When I had first seen the company's own lists of workers suffering from asbestos-related illnesses — given to me by Peter Russell in 1977 — the question had immediately arisen: when would Hardie ever reveal how many people its asbestos had killed? The decades-long delay between exposure to asbestos dust and the onset of disease made argument over the number of deaths and the causal connection with asbestos easy. Disagreement was also common between doctors diagnosing asbestosis from X-rays of those still living, where opinions varied widely. More certainty came from long-term studies of the now-deceased, but Hardie discouraged such surveys of its own workforce, arguing that the high migrant turnover made the task impossible.

The scale of the damage and death caused by asbestos exposure became more difficult to conceal after a direct link between the rare cancer mesothelioma and asbestos was clearly established by the early 1960s, when Hardie then began asserting that the *type* of asbestos was critical.

A quest for the 'safe' level of exposure to asbestos dust generated further argument, and spawned a raft of public health surveys and studies by industrial hygienists. Hardie and its corporate colleagues claimed that the risk of disease from asbestos exposure was insignificant below certain dust levels. The company borrowed many of the tactics employed by its partners in the international asbestos cartel to control Australian research on asbestos and to discredit professionals who sounded the alarm. Backed by a global network of industry PR and research, Hardie dominated key institutions in the public health bureaucracy, both state and federal, and influenced the crucial tasks of setting safe standards and monitoring asbestos dust and disease.

A glimpse of the extent to which Hardie was able to shape views within the bureaucracy emerged from the NSW Dust Diseases Board (DDB), which the government had established as a joint union–industry compensation authority funded by an industry levy.[2] In 1972 Hardie sought details of its former employees whom the DDB had classified as suffering from asbestos-related disease.

To maintain client confidentiality and prevent future discrimination, the DDB decided that Hardie should not be given the names of individual ex-employees. A senior staff member also offered another reason: 'If the names were made available to employers it would be necessary to be even-handed and make them available to (a) the media (b) the environmentalists and there would likely be scare stories put out about the deadly effects of minerals which *industry and the community both need and cannot get along without* [author's italics].'[3]

This assumption that asbestos was an essential product

showed how completely some at the compensation authority had been captured by industry spin. Alternative and safer products had always been available. Indeed, within seven years Hardie itself announced that it was phasing out asbestos.

Not all state authorities were so accommodating. As early as 1947 Dr Douglas Shiels of Victoria's Department of Health had criticised government inaction throughout the world over asbestos hazards. He told factory inspectors: 'The first recorded case of asbestosis was in 1900 ... You would think that as a result of that that the danger of asbestosis was immediately recognised and steps taken to prevent damage by it, but that was not the case.'[4]

Shiels was later to single out the asbestos industry as 'one of the most dangerous' after finding nine cases of asbestosis among about fifty asbestos workers in a survey of dusty trades.[5] He expressed concern that 'many workers' could be misdiagnosed through ignorance of the disease among general practitioners, and asked Dr Gordon Thomas from the department's industrial hygiene division to conduct a more specific survey of asbestos workers.

Thomas examined 300 workers and found positive X-ray evidence of asbestosis among forty-seven. In his report published in the *Medical Journal of Australia* he also warned about the difficulty of getting a complete picture of the number of people affected: '... there must be many more older folk suffering from the complaint in whom it has not been diagnosed.'[6]

Although those at Hardie were not surprised by any of Dr Thomas's findings ('I do not think there is anything in this which we do not already know,' wrote a senior executive[7]), they became alarmed by what was later described as the 'critical and uncompromising attitude' of Thomas and his colleagues.[8] Jock Reid, then Hardie's deputy chairman, raised the activities of the Victorian health officials in conversation with his counterpart at the UK asbestos giant Turner & Newall on a visit to its Manchester headquarters.[9] When Reid forwarded Thomas's study on 'the supposed danger of asbestosis', the UK company replied with similar scepticism: '... to Dr Thomas, asbestosis presents a far greater hazard than it is now considered to present in this country in the light of our accumulated experience ...'[10]

Hardie soothed the concerns of the Victorian Department of Health by installing machinery at its Brooklyn factory to reduce asbestos dust levels. But its international partner in the global cartel fared less well. Turner & Newall was soon at loggerheads with the department over its 'limpet spray' technology, licensed to Australian firms to fireproof buildings by spraying their girders and ceilings with asbestos insulation. Victorian unions threatened to ban the process, and the British company became outraged when Thomas called asbestos a 'killer' while addressing the spray workers. 'If he could save one man's death by asbestosis, his life would not have been lived in vain,' the local company representative tersely reported Thomas had told the workers.[11]

Turner & Newall began a campaign to discredit Thomas, describing him as an 'emotional' man with a 'mission'. Writing from the UK, Turner & Newall's Dr John Knox canvassed several options to reduce Thomas' credibility: 'It could be maintained that the carcinogenic activity of asbestos is not universally accepted ... [and] although some men in Australia are considered to have asbestosis ... no one has yet died to prove it as far as I know.'[12]

* * *

The firm's UK management worried that the 'trouble' with discrediting Thomas was '... that one could not campaign against him without campaigning against the Department, which would be out of the question ...'[13]

The UK company's Australian agent decided instead on a luncheon strategy to marginalise Thomas, inviting the new Victorian Health Director and senior officials from NSW whom Turner & Newall hoped would have more 'intelligent medical opinions'. He reported back to the UK that, following the lunch, the Victorian Health Director had 'no doubt in his mind that Dr Thomas is acting unwisely, to put it mildly'. In fact, the strategy almost backfired when, only moments before the lunch, a NSW test of the asbestos spray process under investigation had resulted in a reading twelve times the recommended 'safe' standard for dust exposure. The agent hastily organised for NSW Department of Health to conduct another test at which '... naturally, every care will be taken to ensure that the results are satisfactory to us'.[14]

To avoid any more 'disastrous' readings that 'would be balm to the soul of Dr Thomas', Turner & Newall's Australian licensees were instructed to 'stall off' any future spot checks until a company representative could attend. And to counter further objections to asbestos spraying by Victoria's 'two Communist-led Unions', the company's agent met with the secretary of the Australian Council of Trade Unions (ACTU), Albert Monk. In a report back to the UK headquarters, he wrote: 'Mr Monk admitted the sinister left-wing union motives and stressed that the right-wing groups were anxious to keep the matter under reasonable control with their preponderant voting power.'[15]

While Turner & Newall was still trying to marginalise Thomas, Hardie, by contrast, was confident that its efforts to cooperate with Thomas in reducing dust at Brooklyn had paid off, and that the Victorian Department of Health would take its advice on 'most matters', including the drafting of regulations.[16]

Indeed, the department had identified another asbestos danger on the docks which fell under federal rather than state jurisdiction. Hardie's factory manager at Brooklyn warned the company's head office in 1959: 'I should point out that the Department is not at all worried about the handling of Asbestos in our Factory, but that they are concerned very greatly with the conditions in the holds of ships carrying Asbestos, where they claim wharf labourers are subjected to considerable dust hazard.'[17]

To feed the voracious appetite of its asbestos factories, the company shipped in millions of bags of raw fibre from Canada, South Africa and Rhodesia (now Zimbabwe). Asbestos bound for Hardie's factories leaked all through the ships. Wharfies would grab the hessian bags full of asbestos out of an airless hold, heave them onto their backs with metal hooks or load them onto a crane, then carry them to the docks, covered in the dust that trickled from the bags. Asbestos might be stacked almost anywhere on a ship, not just in the main hold or between decks. Jim Donovan, later a maritime union official, dreaded unloading it from the freezer lockers, where meat and fruit were stored on the return trips:

> I always found the lockers the worst. If you were down the
> bottom it was a pretty big wide open space, but in a locker
> there was no escape, it was more claustrophobic — there
> were more fibres in the lockers than in the big hold. You'd get
> asbestos all over you all the time. Sometimes the bag would

hook into something going up, particularly if you got a bit of a swing on, and a heap of it would come down on you.[18]

The ports were the responsibility of the federal Department of Health, and Hardie executives had already cultivated a good working relationship with the federal advisory body, the National Health and Medical Research Council (NH&MRC). In particular, the company's close ties with Gersh Major — the industrial hygienist from the department's School of Public Health and Tropical Medicine, whose comment had first sparked my interest in asbestos — were to prove pivotal.

Despite the early concerns about conditions on the wharves expressed by state health authorities, the maritime union was slow to act on the dangers of asbestos. When its British colleagues sounded the alarm in the late 1960s, though, the Australian union threatened industrial action.

Major was called in by the waterfront employers to measure the dust exposure of the workers. He conducted one atmospheric test of the dust in the hold of a ship unloading asbestos in Sydney, where he found an extremely high forty-five fibres per cubic centimetre, more than ten times the so-called safety standard for a working lifetime of eight-hour shifts. Major's boss, Professor David Ferguson, was reassuring when he reported these findings: '... a health hazard is unlikely to be presented to waterside workers engaged in unloading this material from ships.'[19]

Complaints about the conditions on the docks continued to be made by the waterfront union, the employers and WA Department of Health. Hardie's Dr McCullagh and other executives even attended the unloading of a Sydney ship to assess the danger for themselves, and concluded: 'It was agreed that there was no appreciable hazard either for wharf labourers or the general public, but that it was most undesirable that the matter should become an issue of industrial dispute and public comment.'[20]

The Waterside Workers' Federation, unconvinced by Ferguson's reassurances, threatened industrial action if conditions were not improved and regulated. Again the industry sought Major's assistance. He drew up a recommended 'code of practice for handling consignments of asbestos fibre in Australian ports', copied directly from a template drawn up by the British industry's Asbestosis Research Council. The introduction was written to allay union fears:

> As long as there is any airborne dust in the work
> environment there may be some small risk to the health of
> persons exposed to the dust. Nevertheless, exposure up to
> certain limits can be tolerated for a lifetime without
> incurring undue risks. Although there is no recorded case of
> a waterside worker contracting asbestosis from his work, it
> is appreciated that apprehension exists among those who are
> engaged from time to time in such work. In order to reassure
> them and to eliminate any risk, however slight, the National
> Health and Medical Research Council, on the
> recommendation of the Occupational Health Committee, has
> issued a code of practice for the packaging and handling of
> asbestos shipments in Australia.[21]

While the code discussed asbestosis, it made no reference to mesothelioma, the fatal cancer that could result from extremely low dust exposures and which dominated scientific discussions on asbestos by 1972. It concluded:

Some asbestos shipments are packed in permeable bags and stowed in refrigeration lockers, deep tanks and similar enclosed spaces where the concentration of asbestos dust often reaches relatively high values. Occasional exposure to such dust concentrations would not be harmful, but it would be prudent for any man to wear a respirator in these places when undertaking work in such places. For men who find personal difficulties in using respirators, it is suggested that the maximum period for working in such places, undertaken without the use of a respirator, is five days in any one year.

Some obvious questions went unanswered, such as why exposure to the dust 'would not be harmful', but at the same time it would be 'prudent' for a worker to wear a respirator?

The code was designed to cover workers employed handling asbestos in the ports. The truck drivers, however, who delivered the bags of raw fibre to Hardie factories fell under state jurisdiction. NSW Health's Dr Eva Francis measured exposure levels of the drivers who transported the asbestos bags from Sydney's Darling Harbour to Hardie's Camellia factory. She found unexpectedly high dust levels of ten fibres per cubic centimetre, more than double the standard of the day. The levels were especially high when the weather was bad and the bags were loaded indoors. 'They were mainly porous hessian bags,' she later described. 'They were using hooks to grab the bags and there were a great many damaged bags. It was pretty obvious this was going to be quite a problem. They were heaped in sheds on the wharf. Occasionally it was raining and so the truck would come right inside the shed.'[22]

Francis explained that Hardie did not insist that loose asbestos remaining on the back of the trucks was vacuumed up after the bags were unloaded. Instead, the dust was left to billow into the city air as the drivers belted back to the docks for another load, posing a potential danger to passers-by along Parramatta Road, one of Sydney's busiest thoroughfares.

The body counts would later tell the story. Many of these truck drivers were later to die as a result of their asbestos exposure.[23] Nor did the NH&MRC code provide adequate protection for waterfront workers. Despite its promise to 'eliminate any risk, however slight', more than a thousand wharfies were to die from their exposure to asbestos in Australian ports.

Major's code for the waterfront was the first of several asbestos initiatives involving the NH&MRC, which decided to set up a Working Party on Asbestos in 1972 and produce model regulations for the states. Joining Major on the initial national committee was Hardie's Dr McCullagh, Dr Gordon Smith representing the federal health department and Trevor Jones, the NSW health department's chief scientific officer.

Jones, as the only official from a state department, was regarded as sensitive to industry concerns: McCullagh described him as 'a sound, sensible fellow' and Turner & Newall had selected him as one of the 'better informed' officials to attend its lunch intended to isolate his Victorian colleague, Gordon Thomas.[24]

As the senior NSW industrial hygienist, Jones often warned Hardie before he arrived to inspect and measure dust levels in its factories. By contrast, his junior officer, Eva Francis, would launch unannounced inspections, which Hardie countered by conducting its own 'parallel sampling' and questioning her results. She recalled, 'I'd put the dust sampler on one shoulder, they'd put it on the other shoulder. It was a bit intimidating, and certainly unusual. It didn't happen at any other workplace that I ever went to.'[25]

When I asked her on my radio program whether health department inspectors could be easily co-opted by the industry, Francis told me cautiously '… we do have a lot in common with management and it's very difficult, perhaps, to keep an independent stance' — words that Jones later castigated her for using, suggesting that they cast aspersions on the department's integrity. Jones left the department in 1980 and went to work full time for James Hardie.

McCullagh drew up the minutes of the first working party meeting, circulating them first to Hardie's managers throughout the country for comment. His next step was to check them with Jones, before presenting them to the other committee members for discussion prior the next meeting. The working party developed draft codes and regulations for the states to adopt, in a similar manner — Hardie would initiate the first draft, which would then be discussed by the other committee members before adoption by the NH&MRC and the states. The working party also spawned a variety of committees, such as a subcommittee on occupational carcinogens which recommended a safe level of exposure for asbestos.

An NH&MRC code for the handling of asbestos for 'small users' — in other words, tradesmen, contractors and domestic users of asbestos cement and a range of other products, like insulation blocks and woven asbestos — was subsequently drafted by McCullagh, Major and others.[26] Work with Hardie's most popular product, asbestos cement, could be 'carried out safely' with power drilling, said the code, and 'Occasional power sawing does not present a problem if limited to an operation of, say, 15 minutes in a shift, and if carried out in the open air.' (These words drew strong condemnation from NSW Health officials and the code was subsequently amended.)[27]

Major and McCullagh's attention had also turned to developing a more accurate measurement of asbestos dust. Keen to ensure industry control over the process, in 1972 they formed a committee made up of its industrial hygienists, which agreed that: '... top management of Wunderlich, CSR and Hardie should be approached for support and financial assistance ... Only after a preliminary/standardised method is developed and proved as viable, should the State Health Department be asked to comment or participate.'[28]

Five years later, partly through the work of this committee, Hardie developed a more efficient portable asbestos dust sampler that could be attached to a worker's shoulder to take measurements.

Reporting back from a meeting of the International Asbestos Associations in Germany, Major said that Australia was 'far more experienced than most other countries' in measuring the dust. Although a public servant, he had attended the meeting as a representative of the recently established industry-funded Asbestos Association of Australia. James Hardie had paid for his air ticket.[29]

More accurate dust counts were of little use if there was no safe level of exposure. The fatal mesothelioma cancers were occurring in people who had minimal exposure, a fact the newspapers reported with alarm. And a lively debate still raged in the scientific community about the relative danger levels of blue, brown and white asbestos.

To address the increasing death toll, Major proposed an Australian Mesothelioma Surveillance Program. Financed initially by NSW Health and grants from James Hardie and Wunderlich, the planning group included the School of Public Health and Tropical Medicine's Professor Ferguson and Major, Hardie's McCullagh, and Dr Julian Lee, the employer representative on the Dust Diseases Board's medical panel. They were later joined by Major's colleague, Alan Rogers, who had previously worked for British Tobacco. The group had hoped to use its 'extensive old boy network' to provide the data, but this proved inadequate.[30]

The purpose of the program was to keep a running tally of cases of mesothelioma and whether there had been asbestos exposure, or, as Rogers later put it, to '... make scientific sense of a field that is sometimes subject to political activism and emotional outcry'.

Six years after the register began Rogers reported that '... there is a considerable proportion (30–50%) of mesotheliomas in our community that are not due to occupational asbestos exposure. A similar trend is now being reported from many other international mesothelioma studies.'[31]

The suggestion that up to half of all mesothelioma had no association with asbestos proved impossible to sustain. Within a year, the Register's secretary downgraded the estimate, writing that in '... 25–30% of cases no history of asbestos exposure has been found when an in-depth occupational and environmental history has been obtained'.[32]

Whether or not a person with mesothelioma had been exposed to asbestos was the critical issue. The procedure was for an interviewer to gather relevant details from employment and medical records and the patient (or the patient's family or friends) and list these on a form. Two occupational hygienists with 'extensive experience, particularly in asbestos processes', reviewed the forms and categorised asbestos exposure for each case as 'probable', 'possible' and 'unlikely'.

Handwritten comments on the Asbestos Exposure Records[33] reveal that Major, one of the reviewing hygienists, often had trouble believing the case histories:

'Are Italians especially susceptible to mesothelioma? Not a good history and I suspect the part for 1967–1972 as bogus.'

'I think this history was cooked. Too much information from the relative.'

'All the mention of asbestos is not relevant and forced by the interviews.'

'History isn't much use ... because of the experience in other railway workshops. The wife is most likely wrong.'

'This just goes to show that you can never get away from asbestos. One day in 1978 removing lagging. What would [the US professor] Dr Selikoff say about that?'

Major's derogatory comments on the case histories underlined the unreliability of the estimates reported by Rogers and Ferguson. A later study of the lung tissue of cases categorised as having no asbestos exposure under the Australian Mesothelioma Surveillance Program discovered that over 80 per cent had a significant presence of asbestos in their lungs, suggesting that in almost all cases the person *had* been exposed.[34]

A second register set up in Western Australia to track the alarmingly high incidence of mesotheliomas from CSR's Wittenoom blue asbestos mine also found a much lower rate of mesothelioma where no asbestos exposure was apparent. According to Professor Bill Musk of the Sir Charles Gairdner Hospital, the discrepancy between the two registers could be explained by the thoroughness with which the case histories had been taken:

> There's no question the further you delve, the more likely you are to find some history of asbestos exposure. The number of cases attributed to 'no known exposure' varies with the person doing the work and their diligence in making people think back to when it might have happened. It shows up on the graph. I can tell you who was responsible for collecting that data at any particular time.[35]

In fact, the non-asbestos mesothelioma rate is most likely very small indeed, given the widespread use of asbestos and the capacity of the lungs to clear some of the telltale fibres even after the carcinogenic damage may have been done. The latest scientific consensus is that this rate is probably far less than one case per million per year.[36]

With a change of federal government to the Labor Party, the Mesothelioma Program was transferred to a new occupational health authority, WorkSafe Australia. An epidemiologist and occupational respiratory physician, Dr Jim Leigh, took over the running of a revised mesothelioma register: Ferguson, Major, Rogers and Hardie's McCullagh would find themselves on the outer. Leigh began to publish alarming predictions that Australia would not see a peak in asbestos disease until after 2010. He became a particular target for abuse from the 'old boys' and industry.

An address by Leigh to an asbestos symposium in 1996 attracted criticism from the Australian Chamber of Commerce and Industry, which objected to his estimates as summarised in the outcomes paper: 'Australia has the highest known incidence rate of mesothelioma in the world, [and] without effective intervention approximately 10,000 new cases of mesothelioma and at least 30,000 new cases of other asbestos-related diseases have been predicted by the year 2020.'[37] He predicted that the epidemic of asbestos disease would not peak before 2010.

Leigh's projections have to date proven accurate, but the attacks continued. WorkCover NSW complained the following year '... that the data by Leigh ... have long been regarded by actuarial and insurance experts as a gross overestimation'.[38] It confidently quoted a study predicting a peak of Australian asbestos diseases by 2000, a pattern it claimed was confirmed by 'a distinct plateauing' in mesothelioma cases then before the Dust Diseases Board.

Industry projections persistently forecast a peak for the diseases that was too optimistic. Hardie's consultants were asserting in 1980 that a peak would occur at the end of the decade,[39] but by 2006 the company's actuaries KPMG forecast that it would probably not occur before 2010.[40] Yet a subsequent study by the National Centre for Epidemiology suggested that even the KPMG figure underestimated the number of cases by one-third and that the peak could be as late as 2017.[41]

Such repeated failure to correctly forecast the disease burden did nothing to moderate Hardie's spin or mute its attacks on the integrity of scientists who were critical of the industry. When the company moved offshore in 2001, the failure of its actuaries to correctly predict the number of future victims led to the imminent bankruptcy of a foundation left to compensate them. Yet when I told a Hardie PR officer at the time that I would be interviewing Jim Leigh for an ABC TV story about the foundation's plight, she immediately suggested that Leigh's work was suspect, referring me instead to members of the 'old boys network': 'We looked into his work prior to the establishment of the foundation and I recall Jim's views were highly disputed by fellow scientists. Two of note are the Australians Julian Lee and Alan Rogers.'[42]

Privacy laws took their toll on the Australian Mesothelioma Register during the late 1990s. Its major reporting agencies, the State Cancer Registries, scaled back their voluntary notifications of new cases, leaving the WA register and the NSW Dust Diseases Board as its primary sources on asbestos exposure. The Australian Safety and Compensation Council proposed in 2007 to publish this skeleton of data along with details of new mesothelioma cases gathered from the National Cancer Statistics Clearinghouse and the National Mortality Database.[43]

Lung cancer from smoking further complicates the asbestos disease picture. While there appears to be no direct connection between mesothelioma and smoking, there is a strong association between smoking and another cancer of the lung caused by asbestos, bronchogenic carcinoma. Smoking is believed to be bronchogenic carcinoma's biggest cause, and the risk multiplies when asbestos exposure is combined with smoking.

Experts are still debating the level of asbestos exposure needed to develop this lung cancer. The industry argues that before the disease can develop, asbestos exposure must be sufficient to cause the scarring of the lungs that is the hallmark of asbestosis.

Other independent scientists argue that it is not a prerequisite to have developed asbestosis before contracting asbestos-related lung cancer: the asbestos-based lung cancer can occur on its own.[44] These scientists maintain that when people's lungs are exposed to a sufficient quantity of asbestos fibres they face a greater risk of developing cancer. The compensation courts have generally upheld the independent view, though the NSW Dust Diseases Board has continued to spend many millions of dollars unsuccessfully defending the industry position.[45]

On one matter all parties agree. Asbestos-related lung cancers among smokers are significantly under-reported, because doctors seeking a cause tend to look no further than the smoking. German and other studies indicate that asbestos has caused significantly more lung cancers than mesotheliomas, yet in Australia only one-tenth as many are recorded.[46] According to Professor Bruce Robinson, of Perth's Sir Charles Gairdner Hospital, 'There are probably about as many [asbestos-related] lung cancers as there are mesotheliomas, but because we all know that smoking causes lung cancer and that's always assumed to be the only cause, they're hidden, if you like, behind the secret door which we call smoking.'[47]

The true extent of Australian lung cancer deaths caused by asbestos remains unknown. NSW Health's Eva Francis warned her director in 1978: 'We do not have the mechanism to follow up these people [asbestos workers] for recording the incidence of lung cancer or mesothelioma.'[48] It was no good asking the Dust Diseases Board. A briefing note prepared in 1979 for the NSW Minister for Health stated bluntly: 'We are quite certain that they [the Dust Diseases Board] are not notified of all cases and many will have missed that collection agency.'[49]

Compensation claims provide the basis for most of the statistics. But many people who develop a terminal illness choose not to face the trauma of court proceedings; others simply don't realise they can claim or don't know their condition is asbestos-related. Because not everyone claims, the figure is always less than the true incidence of disease.

The asbestos industry recognises this understatement. The more publicity is given to asbestos disease, the more likely asbestos victims are to sue. A surge in compensation claims followed the blaze of media attention on the public inquiry into James Hardie's move offshore in 2001. The company's actuaries speculated that the rise was due to '... increased consumer awareness and association of James Hardie with asbestos, resulting from increased publicity'.[50]

If merely counting the lives lost to mesothelioma was to prove controversial, the argument the industry waged over the *type* of asbestos that caused these deaths was equally intense. And the 'safe' level of exposure to dust in Australia remained in dispute until asbestos, in all its forms, was banned in 2003.[51]

The three main types of asbestos used commercially were blue, brown and white. The first two belong to the amphibole group of minerals and have straighter, sharper fibres than the more commonly used white asbestos which, as a serpentine mineral, has curly fibres.

Blue asbestos was an immediate suspect as the cause of mesothelioma when worldwide concern about the disease first grew from the late 1950s, a danger highlighted in 1960 when the disease was discovered in miners at the blue asbestos Wittenoom mine in Western Australia.[52] Hardie began to abandon its use of the blue fibre in 1968 and soon claimed mesothelioma should no longer be a risk as a result. Hardie's workforce education and public relations material, which it adapted from the British asbestos companies, stated that other types of asbestos rarely, if ever, caused mesothelioma: 'The risks of mesothelioma from the most commonly used types of asbestos — white and amosite [brown] — are believed by informed medical opinion to be negligible, or at any rate very much less than from blue.'[53]

Yet in 1964, Professor Irving Selikoff had found mesotheliomas in American insulation workers whose only exposure was to white and brown asbestos. Indeed, blue asbestos was rarely used in the US. Selikoff warned repeatedly that all the fibres could cause cancer: 'My own experience would not suggest that there is any significant difference between them. Mesothelioma is certainly produced by crocidolite [blue]. It's also caused by the others ...'[54]

Looking back, it seems remarkable that so much time has been spent debating the relative danger of the different asbestos fibres. According to Professor Musk, all types kill and all cause asbestosis, lung cancer and mesothelioma:

> The argument is not worth having, but there is still an argument about it. There is no question that for crocidolite [blue] the risk starts at virtually any exposure. As a general rule I think there should be no exposure: anything above that is a risk. Blue asbestos is the most potent of the fibres. It's of the order of one for chrysotile [white], ten for amosite [brown], one hundred for crocidolite [blue].[55]

But for Hardie the argument was definitely worth having. If blue was blamed for mesothelioma, it could continue to use brown and white, which is exactly what it did. And Hardie had good reason to dispute the dangers of brown. As far back as 1951, Chairman Thyne Reid, while discussing a possible takeover by his UK counterpart, Sir Walker Shepherd from Turner & Newall, had revealed to him a Hardie secret about the usefulness of brown asbestos.

> Mr Reid told me, and emphasised that this information had not been communicated to anybody else, that in their detailed studies of asbestos fibres they have established to their own satisfaction that the best amosite [brown] has as much tensile strength as chrysotile [white], when the relative specific gravities are taken into account. He regards this as a significant discovery which he hopes will enable the considerable expansion in the use of amosite [brown] in asbestos-cement ...[56]

The company was particularly keen to continue using brown asbestos for the same reason that it had retained low-grade white asbestos in the mix for its Hardiflex sheets. Brown was all about maintaining a mix of fibres suitable for its curved products, particularly its profitable pipe manufacturing. At the time Reid confided Hardie's secret to Shepherd, brown made up 15 to 45 per cent of the asbestos mix in the company's asbestos cement products.[57] By 1980, brown comprised nearly a quarter of the asbestos Hardie imported and its 'dominant use was in cement products'.[58]

Evidence of the danger of brown continued to mount through the 1970s. Gersh Major told a court that '... in the 1970s there was a suspected risk of factory workers contracting mesothelioma from amosite [brown] ...',[59] and by 1979 an advisory committee to the British government had proposed a ban on brown similar to that already placed on blue. But Hardie joined the UK industry in lobbying against the ban. Dr McCullagh published a paper in the British *Journal of the Society of Occupational Medicine* claiming that evidence against brown was exaggerated and that there was no justification for treating it any differently from white asbestos.[60] Hardie's Ray Palfreyman continued the claim at a meeting with Australian health officials, putting the position of the South Pacific Asbestos Association, which 'did not go along with the view that a distinction should be made between amosite and chrysotile'.[61]

In Australia, the NH&MRC asbestos ad hoc subcommittee followed the British lead by proposing a ban on brown asbestos in 1982. It noted '... medical evidence that amosite [brown] was substantially lower than crocidolite [blue] but somewhat greater than chrysotile [white] in its potential for causation of mesothelioma and agreed that a hygiene standard lower than that for chrysotile [white] should, ideally, be recommended.'[62]

In other words, because brown appeared to cause more mesothelioma than white, it should have a lower standard. But Hardie would have none of it. As observed by the NH&MRC subcommittee, which included, of course, Gersh Major and a Hardie representative, there was only one known Australian manufacturer still using amosite for asbestos cement products:

The manufacturer indicated that, if a lower standard were introduced, his industry would be forced to seriously consider the exclusion of amosite in building materials. In the case of asbestos cement water pipes no economically viable substitute for amosite other than crocidolite is available nor can one be foreseen at the present time.[63]

The NH&MRC buckled: Australia's premier health research agency left the standard for brown unchanged. It made no assessment of the number of lives that would be lost because of this decision. Hardie continued to put brown asbestos into its products for another three years.

London University's Professor Julian Peto described the failure to ban brown as a 'tragic missed opportunity', resulting in a UK mesothelioma incidence five times that of the US, killing about one in ten British carpenters, with an equivalent rate expected in Australia.[64]

We will never know the full death toll. Plenty of people foresaw that the real incidence of asbestos disease would be far higher than the haphazard Australian statistics would reveal. Hardie's Dr McCullagh's comment that there was no point gathering 'dubious statistics of death' was correct: those statistics which have been gathered are certainly substantial underestimates. Had more accurate statistics been gathered, there is little doubt that Hardie would have been compelled to abandon asbestos far sooner than it did.

five

THE BARYULGIL
'TIME BOMB'

When I first arrived at Baryulgil in 1977 I had no idea what to expect. Nestled on a ridge below heavily timbered mountains, and surrounded by lush sheep and cattle country that sweeps down to the Clarence River, Baryulgil was a tiny mining community near Grafton in NSW. James Hardie had run the nearby mine for more than thirty years and had recently sold it to Woodsreef, another asbestos miner. I had been told that the NSW Department of Health was checking the health of the miners.

As I drove into Baryulgil Square, the fenced area where the community lived, I noticed that the dirt road had changed from red topsoil to a silvery white colour. I picked up a handful of dust and asbestos fibres glinted in the sunlight. I later learnt that the shire council's custom was to line its roads with tailings from the mine. Locals spoke of this practice approvingly: the road was easier to see at night because of its distinctive white dust, and the mine manager had boasted that it was 'the cheapest road-making material on the coast available for councils'.[1]

A few minutes after I arrived a group of families and friends gathered outside one of their fibro homes. They were worried. The previous year their friend Andy Donnelly, a picture of robust health, had suddenly died. As early as 1960 an inspection by the Department of Mines triggered by the Grafton branch of the Australian Labor Party had suggested it might be a 'wise precaution' to X-ray Donnelly for asbestosis, but there is no evidence that it was done, nor that he was ever alerted.[2] An autopsy revealed that his lungs were riddled with asbestos. Until then, the Baryulgil community had regarded the asbestos mine as a handy source of local employment. Now they were not so sure.

Although I had come to interview them — not the other way round — I was immediately peppered with questions from this small group of Aboriginal people. It became obvious that nobody had ever explained the hazards of asbestos clearly to them. 'What have you been told?' I asked.

KEN GORDON: We were told nothing.

PAULINE GORDON: Nothing!

KEN GORDON: Nothing at all. Only the bits you told us today, that's all. That's all we know …

MATT PEACOCK: You've never been told anything at the mine about the fact that asbestos might be dangerous?

DON WILSON: No, never.

NEIL WALKER: I've worked there for nearly twenty years, and I've never been told that.

PAULINE GORDON: If we'd have known, they'd never have been working up there in the first place … We wouldn't have allowed our men to work there. We'd have got away from it.[3]

And so it went on. The group knew a little about the disease of asbestosis, partly from questions they had recently asked the health department staff. They were adamant they would never have allowed their kids to play in asbestos tailings, or 'shivers', if they had known it was dangerous. As we talked I was conscious of dry, rasping coughs in the background.

In 1971 Hardie's Dr McCullagh had expressed alarm at the dust count in the Baryulgil mill of nearly 2000 fibres per millilitre, 500 times the recommended standard of the day.

'With counts of this order the fact that an operator is only so exposed for about an hour a day provides only grossly inadequate protection,' McCullagh had protested at a meeting of Hardie's Environmental Control Committee, and urged the company to spend more money on dust control.[4] His repeated exhortations appeared to fall on deaf ears at Asbestos House in Sydney. The company doctor even likened the dangerous Baryulgil conditions to a 'time bomb ... with the length of the fuse the only unknown'.[5]

The miners were well aware of the dusty conditions, but they did not know the danger. The asbestos ore was blasted out and shovelled into skips, tipped into the mill crusher, from where the fibres were sucked up from the parent rock and blown into a hopper, then dropped into hessian bags. Generally there was less dust in the open cut because the wind often blew it away. Much of the dust was crushed serpentine ore, which was not as hazardous, but on jobs such as bagging of fibre the dust was pure asbestos.

In the bagging area, former boxing champion Tony Mundine told me, he used a pick handle to ram the fibres down into the hessian sacks and the dust went everywhere. 'I was exposed to it as a four-year-old boy, because Dad was one of the first to work on the mine and we used to go up and help him clean the mill at the weekend. We used to think it was great, playing in it as kids.'[6]

He talked about how, as a child, he would slide down asbestos piles at the mine when he visited his father at lunchtime, and how the miners looked like ghosts when they came home, covered in white dust. Like others from Baryulgil, Mundine had telltale asbestos scarring on his lungs and lived with the worry that it might develop into something worse.

Angus Cave, one of the dozen or so non-Aboriginal people to work there, compared conditions inside the mill to '... the worst fog you could ever drive through. We walked outside and I'm darned if you could tell us apart ... we literally changed colour!'[7]

The worst job, according to Cave, was changing bags on the dust collector:

We didn't wear masks or any protective gear ... [you'd] take a lungful of air, walk in and you've got a stick that long to beat the bag. You'd stand there and beat it, like a carpet, and this is stirring all the dust up. It was like concrete! I'd turn around and I'd be out of breath. I'd shoot outside, and breathe in, go back in, and it would take two or three times to change the bag ... When you'd finished that, you'd walk outside and go down underneath ... I could guarantee you'd be sitting in dust four to five inches thick!

It was this process that alarmed a visiting Hardie engineer as early as 1966, when he described the dust hazard as requiring 'urgent attention'. He took photographs, but 'the photographic processor did not print the sock-cleaning operation, presumably regarding it as blank film, which it most certainly is not.'[8]

A spectre of disease has hung over the Aboriginal community ever since they became aware of the dangers of asbestos. The tragic irony was that without asbestos, the Baryulgil people had stood out as a healthier population than many others. Not only was Tony Mundine a world-class athlete, his son Anthony continued the tradition, becoming the world middleweight boxing champion after a spectacular football career. Only the year before the health alarm sounded, the NSW Health Commission doctor who would screen the community for asbestos disease observed that the mine had helped make the Baryulgil Aboriginal community '... one of the healthiest in NSW in the medical, social, economic and cultural sense'.[9]

Probably the earliest compensation claim from Baryulgil was on behalf of Fred Moy, a transport worker who trucked Baryulgil asbestos to the Grafton railway. Moy's exposure was less than those who worked at the mine and mill, but even simply loading and unloading the hessian bags of asbestos at each end of the journey had been enough to cover him in dust. He had applied for a loan and discovered he had asbestosis after he was sent for a medical examination. The transport company's insurers advised him to seek compensation from the Dust Diseases Board, where (after five heart attacks) he eventually received a pension of $20 a week in 1971. Several years later, and much sicker, Moy tried to claim compensation from his former employer through the courts, but was barred by the statute of limitations. He was reportedly living in a shed at the end of 1978, destitute, alone and sick.[10]

During that first visit I heard that the previous generation of Bundjalung people at Baryulgil had formed a special relationship with the Honorable Edward Ogilvie, the first white settler in the area. Ogilvie had learnt their language and employed some of them while building an extravagant stone castle with an inner courtyard and Florentine marble fountain on a nearby bend in the spectacular Clarence River. His granddaughter, Australian suffragette Jessie Street, remembered swimming with the Aboriginal women among the platypuses and turtles as a child.[11]

In 1918 Ogilvie moved the Aboriginal community from the riverbank to the site of today's Square, about ten kilometres away. A small asbestos mine was started near the Square the same year but soon closed. It was reopened by the manufacturer Wunderlich when asbestos became scarce during the Second World War; James Hardie bought a half-share shortly afterwards.

From the outset the mine owners employed local labour, thus distinguishing the Baryulgil community from most other Aboriginal people in the area by providing them with jobs. The community also gained some security of tenure through a lease to the Square from Ogilvie's successor, Sydney retail tycoon Anthony Hordern. The private tenure allowed Mundine's grandfather, armed with a shotgun, to protect the children when government welfare officers tried to remove them to institutions.[12]

Conditions in the mine were dusty from its inception. According to the records of the local Grafton radiologist, within the first ten years of operation at least two miners, Albert Preece and Harry Mundine, had developed symptoms of asbestosis.[13] The following year, in 1953, Hardie bought out its partner Wunderlich to gain full control of the mine. It is doubtful that Hardie wanted the mine for its ore. Engineers like Neil Gilbert and Warwick Lane snorted with derision at the quality of the Baryulgil fibre, which they claimed was riddled with impurities. But Hardie had a good strategic reason for operating Baryulgil that had little to do with the mine itself, and more to do with keeping down the price of its much larger volume of imported asbestos.

At the time of the Baryulgil purchase Hardie was under pressure from CSR to buy blue asbestos from its Wittenoom mine. The federal government, keen to develop the WA mining industry, swung its support behind CSR, proposing '... to induce Hardie to use Australian asbestos by denying them currency for imports from Canada'.[14]

Hardie fought back, arguing that the blue asbestos from Wittenoom was unsuitable for its needs. The small operation at Baryulgil was, Chairman Thyne Reid told shareholders, '... the only mine in Australia which produces, at an economical cost, a grade of asbestos fibre suitable for use in our products'.[15]

After CSR secured a Tariff Board inquiry to consider a tariff on imported asbestos, Hardie relented and negotiated a deal to purchase CSR's Wittenoom fibre for the next decade. But Hardie kept the Baryulgil mine operating to forestall the threat of a future tariff on imports.

The mine was at best a marginal operation that required substantial investment to bring it up to the still-unsafe standards of Hardie's factories. As the former Hardie PR officer Jim Kelso put it: 'The bloody thing never paid two bob anyway. It was just a nightmare of a thing. It went on and on and on because nobody would make the decision to close it down, which should have been done in about 1960.'[16]

Kelso emerged as Hardie's point-man during the federal parliamentary inquiry into the Baryulgil mine established in

1983. An old school chum of Chairman John Reid from Melbourne's exclusive Scotch College, he had only joined James Hardie that year.

Kelso, a metallurgist, had enrolled his children at Pymble Ladies' College when he moved to Sydney, and at his first PLC speech day was surprised to find that John Reid was chairman of the school's governing council. Reid invited him to join. There Kelso met Hardie's managing director, David Macfarlane, who offered him the job of Hardie's community relations manager. By this time Hardie had begun to phase out the use of asbestos from its building products and was shedding its image as a one-product company.

Despite — or perhaps because of — his lack of knowledge about asbestos and the company's history, Macfarlane wanted Kelso to look after asbestos litigation and to defend the company's record at Baryulgil. Kelso expected it to be a tough job: 'The image of the company was severely damaged. It was bad news. Asbestos, Aborigines ... there was everything wrong with it!'

Kelso proved to be an inspired choice. There were many other people far more expert about the Baryulgil mine inside Hardie, but most had frequently voiced their misgivings about its conditions. Speaking with a faintly amused tone, he told me:

I think they felt that it would be better if I, you know, was a fresh brew. There was a suggestion by somebody that the Board wanted to have somebody who knew nothing about the whole business. They thought that it probably wasn't a very good idea if I was filled up with all the old things, and that's probably why they selected me.

The new recruit was coached for his performance ahead of the inquiry. Hardie's lawyers, managers from factories around the country and Dr McCullagh — about fifteen of the company's top executives — gathered in the boardroom in Asbestos House to rehearse him.

> I sat at one end of the table and they sat at the other end, and they started firing questions at me. And that went on for about three hours. I remember that very well because I was a bit worried about whether I'd be able to perform properly. But it turned out alright. I don't think anybody else wanted the job either ... they were glad to be on the questioning side rather than the questioned.

Hardie executives had already forged strategic connections with the newly elected federal Labor government. In the weeks before the inquiry began, Prime Minister Bob Hawke was the company's guest at the Hardie 1000 car race in Bathurst, an event used to promote the firm's brake lining subsidiary.[17] Kelso had also met Minister for Aboriginal Affairs Clyde Holding at the ALP's federal conference, where he said the minister had told him not to worry about the inquiry: 'It will all come out in the wash.'

The Labor government had agreed to hold the inquiry after lobbying by the Aboriginal Legal Service (ALS) in Sydney, which had claimed in the media that up to seventy Baryulgil miners had died from their asbestos exposure.[18] Kelso already knew that the ALS had little evidence to back up this claim because the ALS lawyer Chris Lawrence had naively written to the Canadian Asbestos Information Centre asking for details of the 'cover-up of asbestos hazards in the U.S. and Canada', pointing out the biggest single problem for his Baryulgil clients was the 'inadequacy of medical evidence'.[19] Clearly Lawrence was new to the brief, because the Canadian centre was a promotional body funded by the asbestos industry whose main purpose was to downplay asbestos hazards. It duly passed on details of his letter to the Australian industry association Hardie had set up.

An observer at the hearing, historian Jock McCulloch later gave me his impression as the inquiry commenced:

> The minute I walked in there I thought, 'This is an uneven fight.' On one side sat James Hardie — four expensive-looking people, all very well dressed, Kelso and his lawyers, with their support staff hovering behind. And on the other sat Chris Lawrence of the Aboriginal Legal Service, and his assistant, files on their knees. It was a really unequal battle.[20]

Before proceedings began the Chairman, Gerry Hand, then a backbencher and member of one of the Labor Party's socialist left factions in Melbourne, had talked tough. He told parliament there had been 'blatant disregard' for the health and welfare of the Baryulgil people.[21]

When the joint-party committee began taking evidence in Sydney, Hardie immediately demanded that the proceedings be held in camera, claiming that public hearings would prejudice civil actions being taken against it by former miners. It was a provocative start.

'All hell broke loose,' the ALS lawyer Lawrence later commented. 'There was yelling and shouting and we all walked out.'[22]

The committee members resolved to continue taking evidence in public, but Hardie persisted in using the same argument with great effect through the remaining proceedings to forestall or prevent material from being discussed. 'At the outset I would like to say that I would like to be as helpful as I possibly can to this inquiry,' Kelso told the chairman. But he had a problem. 'Our insurance policies do not allow us to make any admission of fact, or any admission of anything that has gone on, because if we do that, they could become void.'[23]

Hardie, in other words, had decided to participate in the inquiry purely to rebut the evidence against it. Nobody from the company was authorised to give evidence and nobody would. Indeed, the degree of 'help' offered to the inquiry by the company was even more limited. Kelso was soon to declare that Hardie had mislaid most of the documents for the mine it had operated for more than twenty years. They had vanished. 'Regrettably, the details of most of the dust counts were sent to the Manager,' he told the committee. 'I can only say that they seem to have disappeared. A great deal of it has been destroyed during the normal course of business ...'[24]

Could Kelso then provide the inquiry with a list of documents the company *did* have? He had, after all, promised to be bountiful with information. Back came a one-sentence answer from Kelso: 'I regret to advise that the company has received legal opinion that because of litigation pending against companies in the James Hardie group and the difficulty that the committee faces in controlling the use of the material presented to it, I am unable to comply with that request.'

However, the Aboriginal Legal Service had obtained documents from the most recent Hardie mine manager, Gerry

Burke, who by then was suffering from an asbestos-related disease. Kelso strongly objected to their admission. Neither Burke nor the paperwork was to be trusted, he argued, and it was impossible to confirm that these 'alleged' company documents (which were often on company letterhead) were 'authentic, complete or unadulterated'. Further, even if they were genuine, their use would be unfair. He drew a homespun analogy:

> It is similar to any one of us who has twenty, thirty or even more years ago, written to his girlfriends, maybe perhaps even to his future wife, and made a number of statements about the relationship that exists between them. Then somebody comes along some twenty years later and collects all this correspondence and picks out all the juicy bits and publishes them nationally, which would be a fairly embarrassing thing for most of us.[25]

The 'juicy bits' were indeed embarrassing, but Kelso was undeterred. He told the committee that in all his time with the company (then two years) he had found Hardie to be a 'highly responsible company' that would never contemplate knowingly exposing its workers to dangerous conditions. He had found its business morality to be 'beyond reproach'. He even suggested that Hardie had operated the Baryulgil mine as a sort of charity for Aboriginal people: 'One of the reasons the mine was persevered with ... was the very significant economic and social benefits ... the mine brought to the Baryulgil community ... we were the only employer in that particular area and there was a sensitivity to the whole issue with the employees whom we had there.'[26]

The former Baryulgil miners and their families sat at the back of the inquiry. Kelso later dismissed their chronic coughing as an act of theatre.

> They used to bring down loads of Aborigines from Baryulgil. I don't know what that cost them. Anyway, they'd all be in there and somebody had told them that it was important to have a bit of a cough. At one stage I said to Gerry [Hand, the inquiry Chairman], 'If you can't stop them coughing I'm not going to go on. I'm just going to cease altogether and I'll just wait until there's dead quiet.' But the coughing ... they took somebody at their word, which is a bit of a shame ...

He let the sentence trail away with a shrug.

Kelso maintained at the end of the hearings that Hardie had one overriding defence, which he called the 'unthinkable' question: 'Tell us who is sick?' Without bodies, what was all the fuss about?

He was on safe ground. He knew, as did the Health Commission, that even those families whose relations who had been identified at high risk of disease almost never arranged autopsies, the most effective way to prove their deaths were caused by asbestos. And Hardie's own doctor had observed that the only reason more disease had not surfaced in the Baryulgil workforce was because of the high labour turnover.[27] People worked there occasionally then moved away and were heard from no more.

The industry hammered the argument, to the point of racism. The South Pacific Asbestos Association's Max Austin suggested that the inquiry had been convened because of '... a group of disaffected Aborigines who have pushed their claims with success in the media and have come to believe they have an entitlement of some kind because they worked with a hazardous substance, despite the fact that medical examinations have failed to reveal any evidence to support their claims.'[28]

Journalists who had been covering Baryulgil, however, did not need bodies. They had seen the asbestos tailings around the school and heard the experts explain the dangers of asbestos. They knew that at least some of the ex-miners had developed asbestosis.

Much to Kelso's frustration, the nearby graveyard was all the visual image TV crews required:

> There was always some bugger who'd go down to the
> cemetery and photograph the graves. Lawrence had said they
> had all died of asbestos. I suppose it was a bit unkind, but
> we ... by devious means ... found out that one of them fell
> out of the back of a truck when he was dead drunk, one of
> the Mundines. Another one had fallen over and fractured his
> throat or something (I can't remember all the details), but
> there was only one whom they mentioned on the death
> certificate that he had emphysema.[29]

The death certificate belonged to Cyril Mundine, a miner whose autopsy had revealed asbestosis. It had been a bid at Hardie's 1977 AGM on behalf of his widow Ruby to secure more than the $3000 she received on his death which chairman John Reid had dismissed as 'nonsense'.[30]

When the parliamentary committee reported its findings,[31] they were a grave disappointment to the Baryulgil people, but not to

Hardie. Buried in the report were findings against Hardie for negligence in failing to warn its workforce of asbestos dangers, and against government authorities for failing to supervise mine safety. But remarkably, the committee's recommendations failed even to mention Hardie by name.

To Kelso, the recommendations were all he had hoped for. 'The achievement at [the] Baryulgil [inquiry] was not to get mentioned in the report,' he told me. 'We didn't ever think we would win or anything like that, but in the recommendations and conclusions that were made our name wasn't mentioned. That was the only thing that we thought we got out of the whole thing.'

The committee's recommendations did focus on the asbestos contamination of the Square and mine area which, according to the 'polluter pays' principle, Hardie might be expected to remediate. The company itself appeared to anticipate such a liability when it argued that the tailings came overwhelmingly from 'harmless' serpentine ore and contained insignificant amounts of asbestos, an assertion not borne out by other tests.[32]

But no onus fell on Hardie to clean up its waste. Instead, the committee members took aim at the Aboriginal residents of the Square, urging the NSW government to cancel their lease and replace it with one which, while permitting existing residents to stay, would be forfeited if anyone else took up residence.[33]

The inquiry also recommended that taxpayers foot the bill for a 'vigorous' clean-up and stringent monitoring of the entire area, as well as possible relocation of the school. At a cost of many millions, the federal Department of Aboriginal Affairs (DAA) later built Malabugilmah, a new town about five kilometres away, and encouraged the Baryulgil people to move there. The community soon split between those who relocated and those who stayed. DAA refused to improve services on the Square in an effort to force people to leave. As resident Lindsay Gordon told me, '... our fathers fought for it. We will never leave it for anybody.'[34] But the worry in the back of their minds was that they could be slowly poisoned by the asbestos.

Looking back, Kelso regretted the inquiry proceedings had become so acrimonious. 'It was a really nasty business,' he said. 'If you did it nowadays, you'd say, 'What is the incidence of health?' and then you'd try and work out something you could do with the community. It could have been a very productive thing, but it ended up one nasty session after another.'

The former Hardie executive, though, had nothing but praise for Hand. 'Gerry Hand, I've got to tell you, was a very reasonable guy. Even though he wasn't the party that I'd have been supporting, he was very reasonable and I liked him.'

Back at Baryulgil there was nothing but heartache. The Human Rights and Equal Opportunity Commission reported six years later that 'no effective action had been taken either to monitor or reduce the significant health risk' from tailings around the Square.[35] Nor had the federal government set up a specialist Aboriginal medical service in Grafton as the inquiry had recommended. Grafton Aboriginal Medical Service was eventually established in 1991, but it had no special facilities or asbestos expertise, and to all appearances was no different from similar Aboriginal clinics at places like Kempsey, Casino and Coffs Harbour.[36] In time the land at the Square and mine was also remediated, mostly at taxpayers' expense, with Hardie contributing $100,000.

* * *

The Wittenoom expert Robert Vojakovic, president of the Asbestos Diseases Society, visited the Baryulgil people in 1989 and took up their cause. He later visited Gerry Hand, then Minister for Aboriginal Affairs, but came away bitterly disappointed.

'He tried to tell me there was nobody sick and the Square wasn't contaminated,' he told me.[37]

The previous year, Hand had told ABC Radio he wasn't prepared to 'cop' such criticisms. There had been some 'outrageous claims' made by public interest advocates on behalf of the Baryulgil people, he said, and they had been given 'ample opportunity' to prove them. The parliamentary committee had done 'everything it possibly could, within the restrictions placed on us ... to be fair to all people'.[38]

The minister also defended his government's actions following the inquiry to Charles Moran, who had organised a Baryulgil Ex-Miners Association. 'Much has been done to improve the circumstances of the people of Baryulgil,' he wrote to Moran. 'A new village has been established at Malabugilmah, offering far superior conditions ... in addition, the Government has been providing substantial funds to the Aboriginal Legal Service ...'[39]

The parliamentary committee that Hand had chaired had been asked to assess the adequacy of compensation provisions for those affected by Hardie's asbestos. It concluded that the existing legal remedies were sufficient. But in his response to Moran, Hand conceded that the claims for compensation had been long and protracted, though '... it is a complicated matter that is likely to take some time ...'

In fact, the Aboriginal Legal Service was already embroiled in a series of Baryulgil cases that would reveal the inadequacies of the NSW compensation courts. Hardie adopted a strategy of vigorously defending its interests at every stage of the legal process.

The Sydney-based legal firm Turner Freeman had some early success with a claim by the mine's former maintenance engineer, Bill Hindle, who had developed pleural mesothelioma. Significantly, Hindle remembered handling at Baryulgil hessian bags from Penge in South Africa, where Hardie obtained its own brown asbestos[40]; bags from Wittenoom with 'visible quantities of blue asbestos' were also used.[41] As a result, he sued CSR and its subsidiary as well as James Hardie & Coy Pty Ltd and its Baryulgil subsidiary. Hardie argued that Hindle had developed his mesothelioma from earlier exposure in the Australian Navy, but, although he died before his case was resolved, after a series of appeals the company settled with his widow.

It was a different story for the Aboriginal Legal Service (ALS). The first former Baryulgil miner the ALS represented in the compensation court was Billy Briggs, who had contracted asbestosis after working at the mine for about six years. He was sixty-seven and, for a sick man who could neither read nor write, his case was to become a bewildering and ultimately futile process.

As with Fred Moy, the first legal hurdle Briggs's lawyers faced was the statute of limitations. Because the injury had been sustained more than six years earlier, Briggs had only one year to apply to the court for permission to sue. At this stage the law did not cater for the long time lag between asbestos exposure and the onset of disease.

There was also the question of which company to sue. Since its restructure in 1939 James Hardie Industries, the parent company, was able to shelter behind the 'corporate veil', the legal principle under which a company could not be sued for the debts of its subsidiaries. Even though Asbestos Mines Limited, which ran the Baryulgil mine, was co-owned by James Hardie and Wunderlich at the time of Briggs's employment, as the subsidiary it bore primary responsibility. The claim was further complicated by the subsequent sale of the subsidiary to the owners of the Woodsreef mine, who had renamed it Marlow Mines Pty Ltd, but then gone into receivership themselves.

After a special application, Briggs won the first round by obtaining a release from the statute of limitations to sue the subsidiary. Because of the corporate veil, he was refused permission to sue either of the former parent companies, but the Aboriginal Legal Service elected to confront the issue head-on, and appealed. And even though he won this leg of what was becoming a marathon legal battle, in the end the corporate veil worked to protect Hardie. Briggs had won leave to sue the parents, but lost his action. Even the judge commented on the unfairness of his situation.[42]

After nearly a decade of legal proceedings at a cost of over half a million dollars, the ALS cases came to nothing. Hardie's general counsel Mark Knight, who defended the Briggs case, said a number of other Baryulgil claims failed on the same point: '... they were ill-founded cases, on a number of grounds, not just the health, and also the fact that they were trying to get James Hardie when it was really and properly an Asbestos Mines problem.'[43]

In 1989, Robert Vojakovic enlisted the support of another plaintiff law firm, Slater & Gordon, who organised lung specialist Dr Maurice Joseph to screen those Baryulgil people with potential claims. The newspaper headlines trumpeted MINERS SET TO FLOOD THE COURTS, but no asbestosis was found in over one hundred ex-miners examined. Kelso was quoted in the press: 'People say there is something wrong with them and say it is to do with asbestos, but they are not able to establish with medical evidence that this is the case.'[44]

Medical evidence did emerge of disease among the Baryulgil community, but slowly, as would be expected with asbestos disease. It did not always coincide with the expectations raised by lawyers and the media, and in all likelihood many people went undiagnosed. Low Aboriginal life expectancy also probably obscured the impact of asbestos, with many dying from other causes before the time elapsed for asbestos disease to develop.

There is no certainty of the total number of workers, let alone the community members, who may have been at risk from the Baryulgil asbestos. Hardie's actuaries KPMG estimated a total workforce of about 350, with an average of twenty to twenty-five workers there at any one time, rising to forty at moments of peak production.[45]

It is beyond doubt that the dust levels at the mine were high enough to cause asbestos diseases. Nor was there any question that some ex-miners contracted disease: the dust levels were such that they caused asbestosis in individual workers after relatively short exposures, from the time the mine opened until its closure.[46] By 2007, Hardie had paid compensation to fourteen Baryulgil claimants and the NSW Dust Diseases Board listed twelve ex-miners on its books.[47]

Austopsies, which would have offered conclusive proof of disease incidence, were few and far between. The Health Commission returned to Baryulgil in 1981 to re-examine ex-miners for asbestos disease. It found that sixteen had died since its first survey.[48] Despite its recommendation that postmortems be conducted when ex-miners died, none had been done. Given the pleural plaques and other lung damage the commission had already discovered, its doctor reported he was sure that autopsies would have revealed more disease.

The Health Commission surveys identified a high incidence of chronic and serious bronchitis, affecting nearly three-quarters of the miners, almost twice the level of a control group. Asbestos may have been the cause, the Department of Health concluded, but the condition was not necessarily evidence of asbestosis. The task of identifying telltale changes in the lungs was not made easier by Hardie's refusal to allow access to the miners' X-rays, which McCullagh had earlier picked up from the Grafton hospital.[49]

Hardie claimed any link was a media beat-up. 'We were perturbed that certain sections of the media seemed to imply that bronchitis was linked to asbestosis,' the company wrote to the health minister in 1978. 'Our medical advice is that this is not so.'[50]

Government officials at the time concluded that because any damage to health had already been done, there was little point in continuing periodic health screenings.[51] Their particular concern was the longer term risk of cancer. They recommended that Baryulgil people between twelve and thirty-two years of age be monitored every five years, with those over thirty-two checked every two years. They also recommended further research would be necessary, including '... the amalgamation of all available data on persons employed', a check of death registers in Queensland and NSW, and making 'every effort' to gather data on those still living. There is no evidence that any of these recommendations were ever carried out.

In 2006, thirty years after my first visit to Baryulgil, I returned there and met the new doctor at the Grafton clinic, Ray Jones. When he arrived he had received no briefings or files about asbestos, nor was there any special funding to monitor the incidence of asbestos diseases from Baryulgil. As before, almost no autopsies were done. But Jones was concerned about the high incidence of respiratory disease he had discovered among his patients, and he was convinced that the cancer rate in the Baryulgil community was 'substantially higher' than in other nearby Aboriginal communities.

'These people are getting cancers in their forties that normally you get in your sixties,' he told me.[52]

Jones was particularly worried about a cluster of rare cancers diagnosed in three people in their forties, all of whom had played in the Baryulgil asbestos tailings when they were children. Brett Freeburn, one of the three and a non-smoker, had suffered a collapsed lung in his mid-twenties. Doctors had offered no explanation. He had left Baryulgil after school and moved to Sydney, where he married and had five children, and worked as a security guard for Macquarie University. Then he started to get headaches. At first he took a couple of aspirins, but they got worse. Another doctor thought his ears might be the cause and referred him to a specialist, who operated on one middle ear, then the other. The pain persisted, and he was told that it might be due to his teeth. He had all his molars removed, a difficult procedure at the best of times. But there was still no improvement. The pain was intense. 'It got to the stage where I couldn't stand up. I had to have a sleep every day, just lying down, tired every day,' he told me. ' I thought, 'I've been everywhere, and no one seemed to know what was going on and no one could fix it properly.' I said to my wife Tracey, 'You know, something's wrong, but no one can tell me what."[53]

Freeburn began to get double vision and his right eye appeared to be turning inwards. He visited yet another local doctor, who suggested he might have a tumour. This doctor was right. He had a nasal pharyngeal tumour at the back of his throat. The chances of getting such a cancer in the US are about one in one hundred thousand. Yet in the comparatively tiny community of Baryulgil, two other people of Brett's age — Rose Gordon and Albert Robinson — have each lost an eye as a result of the same cancer. Although laryngeal cancer is recognised internationally as a cancer caused by asbestos, pharyngeal cancer remains on the 'suspicious' list. A recent review of asbestos cancers for the US National Academies concluded that the link with asbestos exposure was 'suggestive', but a dearth of studies meant there was not sufficient evidence to declare a definite association.[54]

In the absence of any long-term medical surveillance of the Baryulgil people it is impossible to say how significant this cancer cluster is. Certainly the early experiences of these Baryulgil children were more extreme than most other communities exposed to asbestos. They breathed and played in the dust for hours each day for the first ten or more years of their lives. But when I raised the question of nasal pharyngeal tumours with the Dust Diseases Board's Dr Tony Johnston, he just shook his head. 'We don't recognise that,' he said.[55]

six

HANDLING THE UNIONS: A 'DELICATE SITUATION'

At Bernie Banton's funeral the newly elected Labor prime minister, Kevin Rudd, fulfilled a deathbed promise to Banton by paying public tribute to the union movement. The next speaker, ACTU secretary Greg Combet, wryly observed that it was typical of Banton to take out a bit of insurance: he had received the same request.

'He didn't tell me he had asked someone else.'

But Banton's praise for trade unions was a recent conversion after nearly two decades of disillusionment. His faith in the movement was rekindled after he discovered he had asbestosis and joined the union campaign against Hardie following its move offshore in 2003.

When he worked at Hardie BI, Banton had become a delegate for the Federated Miscellaneous Workers' Union (the 'Missos'), which covered asbestos workers in NSW, Victoria and South Australia. Banton was critical of the achievements of the union and was particularly indignant about the autocratic style of its general secretary, Ray Gietzelt. He began attending meetings at a local pub with the head delegate for the Missos at Hardie's Camellia factory, Frank Shanahan, who was plotting a challenge to Gietzelt's leadership.[1]

What Banton did not know at the time was that Shanahan was receiving secret assistance from the National Civic Council (NCC), an anti-communist, predominantly Catholic organisation run by Bob Santamaria, which operated the trade union group known as 'the Movement'. The Missos was an attractive prize for the NCC. The large voting block it wielded within both the Australian Council of Trade Unions (ACTU) and the Australian Labor Party (ALP) would later enable Gietzelt to be kingmaker for NSW Premier Neville Wran and ACTU President, later Prime Minister, Bob Hawke.[2]

Shanahan came within a few votes of toppling Gietzelt in a ballot in 1971, but Gietzelt comfortably fought off a second challenge four years later and secured convictions against Shanahan and a fellow Hardie organiser for stealing ballot papers.[3] Gietzelt's victory would have profound consequences on Australia's political history. He persuaded Shanahan to testify to the ALP national executive that a key NCC operative, Brian Harradine, had helped organise the union challenge. By a narrow vote Harradine was expelled from the party and went on to become an independent senator who held the balance of power under Prime Minister John Howard's conservative coalition government.

Although Banton had left Hardie by the time of Shanahan's second unsuccessful challenge, he told me that the way his former fellow delegate was treated was 'terrible'. Shanahan developed mesothelioma from his asbestos exposure at Hardie. Ostracised by his former colleagues, he and his family moved to the country NSW town of Wagga Wagga, where he died a painful death.[4]

This dramatic background could have provided some explanation for the unfriendly attitude of the Missos when I first began researching the asbestos industry in 1977. The union's leadership had been under siege and was understandably paranoid. But after the ABC Radio series was broadcast an official from the Missos called me. Although he refused to give his name, he said that there was more to this subject than I might imagine. I still have a yellowing note of that conversation of thirty years ago. He told me that a union man named Shanahan had recently died from an asbestos-related disease and suggested that I contact the former NSW secretary of the Missos, Doug Howitt.

A few weeks after this anonymous tip-off, I tracked down Howitt's telephone number and called him. He sounded very agitated. A line in my ageing notes reads: 'It's all been swept under the carpet.'

I told Howitt I would get back to him, but other asbestos stories seemed to have priority. Hazards from sprayed asbestos insulation in public buildings and elsewhere deserved attention, without chasing a story that was obviously going to require a lot of digging.

But my suspicions about the federal office of the Missos lingered. They were heightened by a confidential company memo sent to me by Hardie's former safety officer, Peter Russell. It had been written in 1957 by an unnamed Hardie executive following an asbestos inspection of the Camellia plant in Sydney by NSW Health. Under the heading 'Trade Unions', it read: 'This is a delicate situation. Little imagination is required to envisage a full-blown emotional trade union attack on the company. However, with individual handling, based on our knowledge of the trade union personalities involved, we feel confident the matter will be resolved satisfactorily'.[5]

Just what was the 'individual handling' of trade union personalities that led to such confidence? The Australian Workers' Union, which covered Hardie workers in Queensland and Western Australia, rarely caused the company trouble: in the fifty-seven-year history of the Newstead plant in Brisbane, for example, there had never been a strike.[6] If any 'handling' was required it would be of the Missos. And it would most likely be Ray Gietzelt, the general secretary of the federal union, who would have to be handled.

The company was obviously worried about the reaction if it informed the union and its members about asbestos disease. The memo cautioned: 'We have an obligation to inform affected employees. For obvious reasons and because in truth there is no cause for alarm on a mass or individual basis, the utmost discretion will be employed.'[7]

Delicate situation. Alarm. Panic. Utmost discretion. The words kept cropping up in Hardie documents referring to its workers. A later letter by Dr McCullagh reflected a similar sentiment: 'The problem is one that has to be handled with some care. The company employs a large number of people whose mastery of the English language is negligible. Among our group it is impossible to be sure that an attempt to create proper concern will not result in implacable alarm.'[8]

And when McCullagh decided to run sputum tests of the workforce to search for cancer, the personnel officer wrote: 'To get the men to cooperate in sputum tests Dr McCullagh will let it be known that he is conducting a survey into the connection between smoking and lung cancer and jolly the men into taking the tests.'[9]

If the federal headquarters of the Missos felt any alarm about the health of union members all those years ago, it didn't show.

Even in 2008, in the front foyer of its headquarters in Sydney, the union proudly displayed a black-and-white photograph taken by Howitt of its former general secretary. Gietzelt stands, felt hat and suit, engrossed in a conversation with four of his union delegates at the loading dock of a factory. The caption read *General Secretary Ray Gietzelt at the James Hardie plant, Camellia.* Three of the delegates are smoking. A closer examination of the photo reveals that they are all covered with a fine coating of white powder. It lines the brims of their hats. It is flecked across their shoulders. It surrounds their boots. It is asbestos.

Nobody knows how many of the union's members died as a result of their exposure to asbestos, least of all the union. Yet the Missos had coverage of more asbestos workers than any other Australian union. At least one of the men in the photos, Jimmy Braid, a head delegate, developed asbestosis and later died of lung cancer. His successor, Frank Shanahan, died of mesothelioma. The fact that the photograph occupied pride of place in the lobby speaks volumes about the union's attitude. It is likely that several thousand of the union's members have died over the years as a result of their exposure to Hardie's asbestos.[10]

Thirty years after our first conversation I made the journey to Doug Howitt's little bush house near Grafton on the northern NSW coast, where he and his wife had developed a wildlife sanctuary.

A feisty Howitt greeted me at the gate: 'You took long enough!'

Even I had to admit thirty years was a bit excessive to follow up on a story.

Howitt told me that his first major involvement with asbestos had occurred in 1956 when, as NSW branch secretary of the

Missos, he and a union organiser inspected an asbestos factory in the inner-Sydney suburb of Alexandria.

'It was awful. The asbestos was festooned like thick cobwebs throughout and the dust spilled onto the footpath and road. We called in the health inspectors and they closed it down.'[11]

The NSW branch was the largest and in those days Gietzelt was a part-time general secretary of the federal union, working during the day at his father's factory and coming to the union office at night.

Howitt continued to agitate over the asbestos hazard, spearheading a campaign through the NSW Trades Hall to include dust diseases other than silicosis in the compensation laws. Under the headline TO FIGHT THE DUST OF DEATH Howitt expressed alarm over the health effects of asbestosis in the union journal.[12] His efforts eventually paid off and the NSW government amended its legislation to include other dust diseases.

His attention turned to the Hardie plant at Camellia where, after an inspection, he informed the company that he was 'very concerned' about the levels of dust he had seen.[13] Howitt's letter caused considerable consternation at Hardie's head office. It took two months for personnel officer Ted Pysden to draft a reply for his colleagues to scrutinise.

Hardie told Howitt it was about to embark on a program of mechanisation, after which '... the dust problem will have been engineered out of existence'.[14] The promise was never kept. The Camellia plant continued to produce asbestos products in much the same way for the next twenty years until it was shut down; Hardie had decided the cost of automation was too expensive.

The pressure of his job and the sense of isolation from his general secretary took its toll on Howitt, who was not a well man following his war service. He was an artillery observation signaller in the Middle East before his deployment to New Guinea, where he had caught malaria. When the war ended he was discharged as medically unfit, suffering from what is now known as post-traumatic stress syndrome. Howitt told Gietzelt in 1967 that he would have to resign unless he got a less stressful job. A new position of research officer was created for him and Keith Blackwell, a close friend of Gietzelt's, became NSW secretary.

Word about the dangers of asbestos was spreading among other Australian trade unions, largely as a result of the publicity that the US Professor Selikoff's study had received both in the US and the UK. An Australian Railways Union official spotted an article about asbestos in the *Times* in 1968 while travelling in Britain, and he circulated it to colleagues on his return to Australia.[15] The Queensland branch of the Building Workers' Industrial Union (BWIU) also alerted its federal office, which passed on the information about the dangers to its other branches as well as the federal office of the Missos. Gietzelt sent this material to his state branches and an article appeared in the union newspaper under the headline ASBESTOS DUST A KILLER, detailing the Selikoff study and reporting mesothelioma in a woman who had washed her husband's dusty overalls.[16]

The building union kept agitating over the issue, calling on the ACTU to prepare a case for federal and state regulations. In South Australia, metal workers in the BHP shipyards at Whyalla held a four-hour stop-work meeting because of the dangers. BHP assured them their health was not at risk, and banned Saturday overtime as punishment.[17]

As research officer, Howitt continued to raise the alarm. He spoke to the *Sydney Sun* about a study he had read in the

British Medical Journal. Under the headline ASBESTOS KILLER, he repeated the call for the ACTU to prepare a case for legislation, warning that the dangers were not just confined to those who worked with asbestos, but included members of the broader community.[18] He cited examples from the British study about asbestos disease in a policeman on point duty, a clergyman who lived half a mile from an asbestos factory, a woman who washed her husband's overalls, and a woman who lived in a fibro house. Howitt also raised his concerns directly with Hardie, sending the company several articles for comment.

Dr McCullagh ridiculed them to his Hardie colleagues: 'I can readily imagine that the policeman who spent 40 years on point duty might well have developed flat feet and a short temper as a consequence of his occupation, but it is stretching the longbow to see any significance in the single asbestos body found in his basal lung smear.'[19]

Soon after, Howitt inspected the Camellia plant with Shanahan, whom Gietzelt had recently appointed a full-time organiser. McCullagh was there to meet the union officials and talk them through Hardie's newly established medical scheme, but Howitt remained unimpressed. 'They took my X-ray that day, showing what good fellows they were, giving free X-rays. Dr McCullagh tried to put it over me. He said you could eat the asbestos, it was so safe. Fancy a medical person, particularly who's been through all that training, saying that!'[20]

McCullagh, on the other hand, believed he had soothed Howitt's concerns. He reported to his colleagues: 'We invited him out to Camellia, fed him lunch, and showed him what we were doing. We are now all good mates and he has not, so far as I know, gone into print on the subject since.'[21]

But Howitt's subsequent silence was not because he liked what he saw at Camellia. Rather, it was because he had been gagged by his own union. After meeting McCullagh, Howitt had continued his inspection of the factory and what he had seen had shocked him.

We went into the teasing room, where they used to empty
big bags of asbestos into a big oblong funnel. There was a
bloke tipping it in. He had his head in it! He was covered in
asbestos! I virtually accused Hardie of being murderers,
I was so upset. Fancy letting a bloke work there like that!
I said get him an air-line respirator, but they were adamant
that he would not wear it. I said get this man a respirator or
sack him if he refuses. Perhaps I shouldn't have said that,
but I did my block. I was a party to his murder. What could
I do?[22]

On his return to the union office he angrily reported the incident to Gietzelt. Looking back, Howitt thought this was a turning point. Both Gietzelt and Blackwell, his successor as NSW branch secretary, accused him of being alarmist. Howitt says Gietzelt told him to mind his own business — that he and Blackwell would look after Hardie — and from that day forward Howitt was frozen out of any union dealings with the company.

'Gietzelt took me off it once I had a blue with James Hardie. 'You do your research,' he said. I don't think I was too popular with the company. I was never involved in any negotiations with them again. What rankles is that he was making sure I couldn't do anything about it. The blokes thought the company was treating them OK.'

More than thirty years later Howitt remained angry, particularly after reading about the ever-increasing death toll from Hardie asbestos products.

'I've lived for thirty years unable to do anything. It's been a lot of stress ... My trouble with Gietzelt came down to asbestos. He wouldn't discuss it with me. I wanted them to change their mixture, to get rid of the asbestos. That, or close it down.'

There is no doubt that Hardie regarded Gietzelt as an ally on the asbestos issue. Hardie's Ray Palfreyman and Ray Gietzelt were so frequently in touch that they were jokingly referred to within the company as 'the two Rays'. Hardie's own documents revealed that Palfreyman made a point of reporting back to Gietzelt after his frequent overseas trips to attend asbestos industry conferences.[23]

On one occasion, Hardie sought Gietzelt's help in silencing criticism from Frank Roberts, the editor of the Australian Workers' Union's newspaper, over a new asbestos safety standard proposed by the National Health & Medical Research Council. Roberts had visited Professor Selikoff in the US, where he had familiarised himself with the US asbestos regulations, and he had expressed outrage that the NH&MRC had failed to consult unions about its proposed regulations for Australia.

Hardie feared that Roberts would '... precipitate public discussion of the issues involved' when he returned from the US. Hardie's Environmental Control Committee noted: '... these [US] regulations would be far less acceptable to the industry than the NH&MRC proposals ... Mr Palfreyman had already taken steps to head off such a development by making use of his good relationship with Mr Gietzelt of the Miscellaneous Workers' Union ...'[24]

Howitt was not the only branch secretary in the Missos to raise the alarm over asbestos. He had discussed the matter with his Victorian counterpart, Ray Hogan, who was similarly concerned. When the Victorian branch requested a meeting with Hardie in 1967 to discuss the health hazards of asbestos, Hardie again sought Gietzelt's assistance in deferring the talks until it had set up a dust-extraction and medical surveillance scheme for its workforce.[25] Dr McCullagh reported that Gietzelt promised to put aside a union claim for dust money if the company's exhaust system kept the dust down, describing the establishment of the scheme as something 'wonderful'.[26]

When I met Gietzelt in 2005 he played down the dangers of asbestos, and was frankly sceptical about the hazards of asbestos cement.

> A lot of people have gone overboard and called for a ban on asbestos cement. I built my house from fibro. In 1944 and 1945 all I could afford was a fibro home. Over the years I renovated and jacked it up, and sawed and cut the fibro. I knew I had to have a mask on and I had to be careful that some of the dust didn't float around. It's minimal, absolutely minimal risk.[27]

Nor did the former general secretary believe that many of his union members had died as a result of asbestos exposure. He was sure that 'if it had been a lot, I would have heard about it'.

He would not accept Hardie's own latest actuarial estimate of compensation claims from its factories, which itself was an underestimate of the true incidence of disease.

'I can't believe that. I think that's a much exaggerated number. People are dying all the time ... when they got rid of the blue asbestos ... there wouldn't be many of those sort of people around now.'

Gietzelt had just finished writing his memoirs, which are mute about the dangers of asbestos, although he recorded other details of his union career with great clarity. When I asked if he could recall discussions with Doug Howitt about asbestos, his memory was vague. He denied having any arguments with him about it.

I slid the 1958 Hardie memo across the table to Gietzelt, the one that talked of a 'delicate situation' with the unions and speculated on whether key trade union personalities could be 'handled'. How did he think he was handled?

He thought for a moment, then suggested the fact that Hardie paid Shanahan's wages while he worked as a full-time delegate for the union might be a possible answer:

> There weren't too many bloody employers in Australia doing
> that in the fifties. That's something to their credit, but on the
> other hand that could have another motive too, of course ...
> that at least you're neutralising any, you're getting a feel
> about what's happening at the shop level ... I can truthfully
> say, Matt, there was no attempt in any way to try to
> influence me to go soft ... I had a good personal relationship.
> They knew I was a no-nonsense person ... they wouldn't
> have offered me a bribe or anything like that. You've got to
> give them a measure of intelligence, you know.

But the union under Gietzelt's leadership certainly appeared to be 'soft' with Hardie on asbestos. When the Missos first raised safety as an issue in a claim for a disability allowance for asbestos cement workers before the NSW Industrial Commission in 1976, the union lawyer was unusually conciliatory: 'We are not seeking to lay the blame on the employers ... It is our view that so long as society insists upon the need of such commodities as fire-resistant asbestos materials, then surely society must be expected to pay some penalty, some compensation for the hazards and the disability involved in its production.'[28]

The commissioner witnessed the hazards first-hand. At an inspection of Hardie's Camellia factory during the hearing, he was confronted by a figure 'resembling a Yeti' who emerged from under a big metal vat covered from head to foot with asbestos. The man had been cleaning out the blender where raw asbestos was fed into the plant. At the hearing, Hardie's lawyer played down the incident as a freak occurrence, although it sounded suspiciously similar to the scene that Doug Howitt had witnessed at the same spot years before when he accused the company of being 'murderers'.

Gietzelt claimed the union had done what it could about the asbestos hazard. When Hardie opened a high-tech automated factory in the rural NSW town of Moss Vale in 1979, he told the local newspaper that '... the union had negotiated with James Hardie for more than two years to minimise the risk to workers'. Yet despite its modern design the factory often exceeded the recommended asbestos level.[29]

Hardie eventually abandoned asbestos, according to Gietzelt, because of union agitation. There was union agitation, but it did not come from Gietzelt. Rather, it was driven by Ray Hogan, the Victorian branch secretary, who during the 1970s took over the battle against asbestos that Howitt had begun in NSW during the 1960s. Like Howitt, Hogan had come to the view that asbestos was so dangerous that it should be banned altogether, a view Gietzelt told me ignored industrial reality:

> Ray Hogan saw asbestos as a deadly thing where workers
> were dying and their health was being impaired and
> thought that we shouldn't be manufacturing it — we
> should bloody well stop it! Well, you can't do that. You can
> have all the pious ideas in the world about things, but you
> can't just bloody well shut it down and ignore the
> industrial and economic and political realities of a
> company operating.[30]

Hogan, a member of one of the Victorian ALP's socialist left factions, also opposed Gietzelt on uranium mining, an issue that tore the Labor Party apart during the 1970s. Gietzelt joined Prime Minister Hawke in changing ALP policy to allow mining within the Northern Territory's Kakadu National Park with coverage provided by the Missos. Hogan's opposition was further proof of his stupidity, according to Gietzelt.

'I had this first class row with him because he took this stupid, narrow, parochial view that we shouldn't be involved in it ... he was showing weakness in taking up this puritanical view that used to shit me ...'

Hogan lost his battle against uranium mining, but against Hardie he eventually prevailed, although he said his task was made more difficult because of opposition from his own members:

> We sought to get Hardie to stop using asbestos in the
> manufacture of the materials we were making ...

The members really thought that it was as safe as making ice-cream. Like we were a few years earlier, they had no knowledge and were not given any information on what they were dealing with. We had to put out material for them in a number of languages because they were mainly migrant employees, but after a reasonable amount of time they understood that they were putting their lives at risk.[31]

One union member, Theo Meletis, contacted Hogan several years after Meletis had been sacked by Hardie. He had worked in a job similar to the one Howitt had witnessed, loading the raw asbestos into the mill at the Brooklyn plant in Melbourne, and was frequently covered in asbestos dust. Hardie's first round of medical tests on Meletis in 1968 had found pleural thickening consistent with asbestosis.

The Hardie lawyers told Meletis they did not want the bad publicity associated with a common law claim. To coax him into dropping the lawsuit launched by the union on his behalf in favour of a private agreement, Hardie enlisted the support of his priest. After several years of legal skirmishing, Meletis eventually accepted Hardie's offer.[32]

Hogan was outraged by the company's behaviour: 'The Hardie people really approached the matter with a high degree of arrogance ... There is no doubt that the top echelon at Hardie's knew that the asbestos was highly dangerous and a killer and yet they went ahead and continued to use the asbestos. The way they conducted themselves was quite disgusting.'

The union asked Hardie to provide the union members with copies of their medical files, but despite an earlier promise the company refused to provide them.[33] Hogan also sought a special cash payment for workers identified as having asbestos disease to tide them over before they won a compensation claim through court. Hardie relented on both these demands only after a series of strikes.

During the industrial campaign for a compensation payment, Hogan began receiving phone calls from an anonymous Hardie executive he nicknamed 'Idi Amin', who tipped him off about the company's tactics. When the critical union vote took place on Hardie's offer, the Victorian branch opposed it, but the national union vote approved the deal with NSW and South Australian support. 'Idi Amin' warned Hogan before the vote that Gietzelt had already told Hardie's Ray Palfreyman that 'NSW and South Australia are OK'.

Hogan was loathe to criticise Gietzelt when I spoke with him in 2005 at his Melbourne home. He told me quietly, 'I could have criticisms of Gietzelt, but he did a quite magnificent job of turning a little right-wing union into the union it is today.' Nor would Hogan disclose the identity of his Hardie informant, a secret he took with him to his death a few weeks later.

As the dangers of asbestos attracted more publicity and attention in the late 1970s, unions with a far smaller presence at Hardie than the Missos began paying greater attention to the working conditions in its factories. Vic Fitzgerald, an organiser with the tiny Federated Engine Drivers' and Firemen's Association (FEDFA), first visited the Camellia plant in 1978. He was shocked by what he saw.

We went into the storeroom where the asbestos was kept in big hessian bags, a bit like those big bales of paper. It was loose, and the forklifts used to puncture the bags. The asbestos used to float around in the air. It reminded me of those little paperweights, you know, the ones you'd shake and there'd be a snowstorm. I put my hanky over my mouth and said, 'Let's get out of here!' Everyone had white dust over them, on their overalls, in their hair ...[34]

Fitzgerald complained to the Hardie manager.

'That's OK, mate,' he was told. 'Blue asbestos is the one you have to worry about. This is white. It doesn't hurt them. It might give them a bit of a cough, that's all.'

A gruff giant of a man with a dry sense of humour, Fitzgerald had heard my ABC radio programs and was already aware of the dangers from his experience of power stations, where asbestos insulation had cut a swathe of death through the union's membership. He told the manager he might fool the workers, but not him.

'A lot of people came and went, and a lot went with death in their lungs. They were convinced to a man it wasn't hurting them. From the very top down they had people convinced that the story they'd put about was true. But those at the top had to know. They've got a lot to answer for. We'll never know how many people died.'

The FEDFA's delegate at Camellia, Stan Fleming, was 'a lovely little bloke', Fitzgerald recalled, whose job was to grease the machinery. He was soon to die a 'horrible death' from mesothelioma. At his first meeting with union members at the site, Fitzgerald had explained what he knew about the hazards of asbestos. He was soon to experience the duplicity of Hardie management, which decided he could take the blame for a long-overdue, if unpopular, safety measure.

> A couple of days later, Stan rang me and said, 'Get down here quick, these blokes want to hang you!' I got down there and found that the company had stopped the blokes from wearing their overalls home. When the bell rang there'd be ten minutes to wash up, then the company had allowed them to jump in the car and drive home in their overalls. Now they had double lockers installed and they had to change. I told the blokes it was a company decision, but they might thank me one day. Not one ever rang.

From Fleming, Fitzgerald also gained an insight into how the company managed its sick workforce. The workers were convinced that Hardie had their interests at heart.

'Fleming told me, "This company looks after us. They send us off to the Dust Diseases Board to see how we are, and as soon as you reach a certain level they'll put you off."'

The reason for this Hardie strategy was simple. The no-fault DDB process was cheap, fast and secret. The compensation claims were cross-subsidised by other employers, there were no lawyers and Hardie had influence over medical determinations through the employer representative on the board's Medical Authority. Claimants received as little as a sixth of the money they would get if they claimed successfully under common law, something the statute of limitations barred them from doing unless they did so within one year of receiving the DDB pension.

Action taken by another union delegate at Hardie, this time from the manufacturers' union, highlighted the unfairness of this treatment. Tom Cook started with Hardie at Camellia in 1977 as a maintenance fitter. Cook knew nothing about asbestos and, like Fitzgerald, was told that '... the blue asbestos was the dangerous asbestos and white was OK'. He researched the subject and began compiling a list of workers who had become sick from asbestos, in much the same way as Peter Russell had done twenty years before. He had identified seventy-three cases by 1987, describing the toll in the AMWU newspaper as a 'modern-day holocaust': 'At least 36 are dead (the fate of eight is unknown), at least three are 'living on borrowed time with cancer', many are dusted to an extent that has ruined their health, one has had a lung removed and one saw his wife die of anxiety over his condition.'[35]

Cook began advising workers about their health and compensation rights regardless of what union they belonged to. He also visited ex-employees he'd been told were sick at their homes or in hospital and referred those who were ill to the AMWU solicitors, Turner Freeman. For many sick members from the Missos who had simply been processed through the Dust Diseases Board, the referral came too late.

Alf Hinton, one of those referred to Turner Freeman, had worked as factory foreman at Hardie BI in Camellia, by far the most dangerous of the Hardie plants, where the brown asbestos dust was often ankle-deep. He began helping Cook bring in former employees to see the AMWU's law firm. Hinton had asbestosis and later died from peritoneal mesothelioma.

For Turner Freeman's Armando Gardiman, the steady procession of seriously ill workers who came to see him was a tragic eye-opener. For many, he found there was nothing he could do.

I was just staggered. These guys were really, really sick and some of them had been sick for years. They were all 100 per cent asbestotic. They could barely cross the room without being in severe respiratory distress, they were just so severely asbestotic. And they were all statute barred, under the statute of limitations. The whole lot of them ... dozens of them.[36]

In addition to the Dust Diseases Board pension (in other states, a workers' compensation pension), the sick Hardie workers could also claim a lump-sum payment from the special compensation deal negotiated between the Missos and Hardie. But in NSW, the state with the largest number of Hardie employees and home to the Missos' national headquarters, the union was slow to process the members' claims. Gardiman found himself handling a string of their payments himself.

'I wasn't supposed to, it was all supposed to be done through the union, but I used to get the guys to fill in all the forms and I'd attach all the medicals and post them in for them. Then Hardie would process them ... get the bloke in and get him checked, absent the union ... so that was that.'

Twenty years later, Hardie's largest union was missing in action as the rest of the country witnessed an extraordinary trade union campaign to drag James Hardie back from the Netherlands to provide adequate compensation for its asbestos victims. The campaign, initiated by the manufacturers' union, drew in the ACTU and other unions, even mobilising the support of international affiliates. According to Barry Robson, former NSW secretary of the Maritime Union and president of the Asbestos Diseases Foundation of Australia, the Missos were invisible.

'Funny, when you think about it,' he said. 'It's just so obvious it didn't occur to me.'

A year after Hinton began coaxing the sick and dying members of the Missos to visit the AMWU's lawyer, Gietzelt retired as general secretary. That same month Prime Minister Hawke appointed him to the board of the national airline, Qantas. Ironically, the deputy chairman at the time was none other than John Reid, then still chairman of James Hardie Industries.

Gietzelt recalled a night of conviviality when Reid, also active on the Sydney Opera House Trust, invited the board members 'and their wives' to a night at the opera.

'I found him very gentlemanly and quite a good bloke to get on with, and I think he would have shared the same view of me,' he told me.

Did they ever discuss the asbestos health issue?

He thought for a moment. 'I wouldn't have raised that with him, no. It wouldn't have been protocol for me because if I had raised that with him, as a member of the board, it would rather indicate that I'm sort of holding you responsible.'[37]

seven

SPINNING THE REID 'LEGACY'

The man who presided over James Hardie's belated exit from asbestos was John Boyd Reid, who succeeded his father Sir John (Jock) Reid as chairman in 1973. When John junior took control of the family's company Hardie was producing more asbestos products than ever before. At the Camellia plant in Sydney the machines were turning out a kilometre of asbestos cement sheeting every hour.[1] There was no sign that he would soon be forced to abandon the mineral that had made his family so wealthy. But within the decade, the dangers of asbestos became a secret too big to hide.

On his retirement Sir Jock told his shareholders that there had never been so much asbestos going into Australian house construction. Indeed, '... it would be difficult to find a newly constructed house in which some asbestos cement was not used'.[2]

His son was even more enthusiastic. At his first AGM, in 1974, he voiced what now stands an ominous warning to home renovators: 'A careful examination of most houses being constructed today will reveal a greater volume in square metres of our product being used, sometimes in unexpected places, than in past years.'[3]

Such 'unexpected places' included areas like the bathroom floor or the interior walls. And he proudly declared that the company had also developed new uses for asbestos outside the house. The corrugated Super Six roofing (which had previously contained blue asbestos and still contained brown asbestos) was '... enjoying a gratifying growth in domestic, industrial and commercial roofing. Its popularity as free-standing fencing is increasing rapidly in all States.'[4]

The chairman continued to hail the company's marketing success in 1977, proclaiming that asbestos cement was increasingly being used in building applications 'undreamt of ten or even five years ago' as a substitute for other materials.[5] The following year, Reid boasted that '... every time you walk into an office building, a home, a factory; every time you put your foot on the brake, ride in a train, see a bulldozer at work ... the chances are that a product from the James Hardie group of companies has a part in it'.[6] What he didn't add was that he, his fellow directors and Hardie executives were also aware that people who worked in all of those places might develop cancer as a result.

A lawyer by training, Reid began working with the James Hardie parent company in Sydney in 1957 as legal officer, a position that required knowledge of the company's potential legal liabilities. The first asbestos compensation claim had been filed against Hardie nearly twenty years before. Since then dozens of additional asbestosis cases had been identified at the company's factories around the country. Several state health authorities had also expressed grave concern about the asbestos cancer risk.[7]

The young John was raised as a gentleman in a rarefied atmosphere of wealth and power, heir to one of Australia's largest family fortunes. In addition to the Melbourne home, the family had a farm at Launching Place and a seaside house on the Mornington Peninsula. He was educated at Melbourne's exclusive Scotch College and his family was a pillar of the Presbyterian Church. Here he learnt his stiff-upper-lip style, rarely displaying his emotions. In later years, when he called his wife Virginia on the phone, he would invariably greet her with the words 'Reid here'.

Melbourne was then the centre of Australia's business world, and, before taking the chair, Sir Jock ran the group's marketing and sales from Collins Street. As a twenty-year-old undergraduate his son John sometimes manned Hardie's exhibition of asbestos cement homes at the Melbourne Gas Company's showrooms. Sir Jock travelled overseas almost every year and entertained frequently. He had a dining room lavishly fitted out at Hardie's city office for lunches of up to forty people and would fly in guests from around the country. Conservative politicians like Western Australian premier Sir Charles Court and his Victorian counterpart Sir Henry Bolte were frequent visitors. So, too, were executives from the US-based Johns Manville and Britain's Cape Asbestos and Turner & Newall. As vice-chairman of the Australian Broadcasting Commission, Sir Jock often invited senior ABC executives over to his seaside house on weekends.

After his stint as legal officer at Asbestos House in Sydney, John Reid's first assignment in 1960 was as manager of Hardie's South Australian operations where the company had secured a state-wide monopoly over manufactured asbestos after buying out its only competitor, Wunderlich. Its product was in huge demand. To accommodate a flood of British immigrants, tens of thousands of fibro homes were built by the Housing Trust of

South Australia in the outer suburbs of Adelaide like Elizabeth and Salisbury. Hardie also provided asbestos insulation for BHP's Whyalla shipbuilding facilities and pipes for Adelaide's water and sewerage. The asbestos waste from its factories was dumped at local tips and was eagerly sought as cheap fill for driveways and paths.[8]

When his uncle Thyne Reid died at the end of 1964, John began his ascent as the empire's heir apparent, assuming his place on Hardie's board and moving to the Sydney headquarters to work alongside his father, who had taken over as chairman. He was soon made responsible for expanding the company's reach into Asia, travelling to Malaysia and Europe, where the young Reid was introduced to the global asbestos cartel that his grandfather had joined in the 1930s.[9] Reid learnt how the cartel divided up global markets and how it evaded sanctions against the apartheid regime in South Africa, the source of most of the world's brown and blue asbestos. In a letter home he explained how Cape Asbestos, which Hardie part-owned and which supplied its brown asbestos, was importing South African asbestos into Singapore and then freighting it across the Johor-Singapore Causeway to Malaysia to another asbestos manufacturer, Hume: 'Hume's story if they are ever caught is simply that they are bringing in Australian blue across from Singapore. They have an arrangement with Cape to bring it in unbranded bags.'[10]

Conditions in Cape's Transvaal mines were notorious even by the low standards of the times, with young African children trampling raw fibres in huge jute bags, whipped on by burly overseers.[11] Today scarcely a family in the Transvaal has not been stricken by asbestos disease.[12] A landmark action against the company through the British courts in 2003 won a small settlement for 7500 South African miners and their families.[13]

The threat of mesothelioma was being widely discussed in the top echelons of the company by 1969, when John Reid was appointed deputy chairman of James Hardie Asbestos. The family and senior managers at Hardie read widely and were well informed about the latest trends and research in Europe and the UK. But the enveloping storm over asbestos dangers was rarely discussed outside a small circle of senior executives and directors. In keeping with its overseas partners, Hardie's asbestos production continued to grow at a phenomenal rate, with global usage climbing dramatically until the mid-1970s. In fact, the vast bulk of world asbestos production occurred after the dangers of mesothelioma were widely known.[14] Looking back, one senior executive close to Sir Jock cringes at the memory.

'You look at it in retrospect, at what you did yourself ... you enhanced the problem. It's morally, ethically criminal. I have nightmares at some of the things we've done.'[15]

There appear to have been no such qualms within the Reid family. For Sir Jock, it was all about 'stewardship', a subject about which he lectured his associates frequently. It was many years later that the executive realised such 'stewardship' was focused on profit for the family and shareholders rather than the welfare of Hardie's employees or customers.

John Reid played a vital role in expanding Hardie's operations. By 1972 he had been invited to join the board of BHP, a major user of Hardie asbestos. In the same year he oversaw the construction of an asbestos factory in Indonesia, a joint venture with the wealthy Bakrie family, which was opened by the Australian prime minister, Malcolm Fraser.

The following year Reid, as Hardie's chairman, increased the company's share in the Canadian miner Cassiar, where most of its white asbestos was sourced, to 12.5 per cent. He also bought out the UK-based Turner & Newall's stake in their joint Malaysian asbestos operation, raising the company's shareholding to 40 per cent. By 1977 Reid had secured an effective Australia-wide monopoly of asbestos cement manufacturing with the purchase of Wunderlich. The same year he opened a new multi-million-dollar, state-of-the-art asbestos cement pipe factory in semi-rural Moss Vale, in the NSW Southern Highlands, the declared purpose of which was '... to provide more space at the company's Camellia plant for expansion of building products manufacture'.[16] All this expansion was overseen by a chairman who still appeared to believe the company could 'engineer the dust out of existence'. In fact, the custom-built Moss Vale factory was only to last a few short years before it would be abandoned by Reid.[17]

Alongside this growth in asbestos manufacturing, the worldwide industry invested heavily in PR campaigns designed to neutralise negative stories about the mineral's links with cancer. Hardie factory managers were told that the challenge of the seventies was to promote a 'new healthy image' for the fibro house;[18] across the ocean, Cape's US subsidiary published an 'Asbestos Family Hero Album' and discussed other family-oriented promotions suggested by PR consultants, including a children's book featuring 'Asbestos-man' and his helper, 'Smokey the Bear'.[19]

In Britain, the industry established the Asbestos Information Council (AIC), which issued press statements downplaying the hazard, comparing asbestos cancer risks with those of 'charcoal-grilled steak'.[20] Hardie borrowed the AIC's media releases and adapted them for local distribution; home-grown statements were held 'in readiness' by the company's in-house PR department to minimise any 'unfortunate publicity about asbestos-related disease'.[21] With the exception of a 1975 cover story in *The Bulletin* magazine by journalist Timothy Hall, headlined IS THIS KILLER IN YOUR HOUSE?, Australian reporters generally seemed prepared to accept the company's reassurances. Hall wrote a sequel to his story titled A MACABRE WAITING GAME, tracking the trail of death which was by then accumulating from the use of the Wittenoom blue asbestos in items such as welding rods.[22] A copy was circulated to Reid and his fellow directors as well as senior managers, with the note that '… it is unlikely that the Company will reply …'. They knew the Wittenoom blue had been mixed into Hardie's asbestos products. It was, indeed, now a matter of waiting to see the result.

There was not long to wait. Although Hardie's record production levels would later cause even more deaths, by the end of the 1970s disease was already occurring among former customers. A flurry of media interest sparked by my ABC radio *Broadband* series placed the company in the centre of an emerging scandal. Reid was soon to oversee a multi-pronged legal, financial and media corporate strategy to minimise the damage caused to Hardie by the growing Australian epidemic of asbestos disease.

The real catalyst for change at Hardie came in the form of a new managing director, David Macfarlane, who replaced Reid's former mentor Ted Heath in 1978, when Heath's alcoholic excesses became too much even for his chairman to tolerate.[23] Macfarlane had worked with Reid in Adelaide and Malaysia, and he quickly became concerned about the explosion of stories in the media over issues such as asbestos insulation in school roofs.

To help influence the reporting, Macfarlane hired Hill & Knowlton's PR subsidiary, Eric White Associates, whose NSW director Leslie Anderson and colleague Bill Frew began a relationship which would soon rewrite the history of the company, erasing Hardie's association with asbestos from public memory and record. Reid himself was to embrace a personal PR makeover, to be cast in the role of a 'Medici-like' philanthropist,[24] a patron of culture, the arts and medical research. Within two years, Hardie executives were being invited to drink cocktails at Sydney's American National Club with the vice-president of Hill & Knowlton, Richard Cheney, as the fortunes of the two firms intertwined.

Initially Eric White's advice made the case for the 'continued use of asbestos'.[25] Frew recommended that briefings be prepared for sales teams, architects and other major users and retailers to emphasise that '... health problems associated with asbestos are related to heavy exposure to dust over a protracted period ...', a claim both Eric White and Hardie knew by then was a demonstrable falsehood.[26] Under pressure from the NSW Health Commission, Hardie had begun to place what it privately admitted were 'innocuous' warning labels on its asbestos cement sheets, but its advisers warned that even these were likely to raise questions in the minds of its customers and retailers.

The PR line that 'the risk to end-users of Hardie products is negligible' was becoming harder to sustain. A director of Hardie's other boutique PR agency, Neilson McCarthy & Partners, urged a quick edit to one of Hardie's 1978 home handyman TV advertisements which had made him 'wince', because dust was 'quite clearly visible' when the asbestos cement sheet was cut.[27]

Eric White's Bill Frew coached Reid on how to answer any 'awkward questions' that could be raised at the company's 1979 annual general meeting. He urged the chairman to stick closely to the wording of a booklet already prepared for the Hardie workforce:

> Since the beginning of this year James Hardie asbestos cement products have carried a warning label to ensure their sensible use. This is in line with the widespread practice of labelling products which could pose any health risk, no matter how remote ... Asbestos cement products have been used in Australia for 75 years. In that time we are not aware of any case of illness among people whose only contact with asbestos has been by way of AC [asbestos cement] products.[28]

The words were carefully crafted. If there were any risk, it was 'remote'. Hardie was 'not aware'.

Frew was to encounter even trickier semantic problems as the company continued to maintain the myth that its customers were not at risk.[29]

Following the AGM, he wrote to his colleague at Hardie:

> ... there are still one or two questions for which we have not developed adequate answers. The question of the number of people who have contracted asbestos-related diseases in Australia is the major area of weakness as is our inability to quantify the number of James Hardie employees affected.

Another potential problem area related to this is the question of mesothelioma among builders and tradesmen whose only readily identifiable contact with asbestos has been by way of working with asbestos cement.[30]

The health risk to such Hardie customers, it seemed, may not have been quite so remote as Reid had told shareholders. Frew recommended the commissioning of dust tests for handymen working with asbestos cement, an essential step, he advised, '... particularly if there is a real possibility of the now established mesothelioma registry and the Dust Diseases Board turning up cases of disease amongst tradesmen, and worse still, handymen'.

Hardie's PR material continued to proclaim publicly that 'tests have shown' that a person cutting asbestos cement sheets 'would not be exposed to dangerous levels of asbestos dust'.[31] Yet a year later Frew asked: 'Now that our tests have been completed, do they show that a person cutting AC sheet according to the rules, should not be exposed to dangerous levels of dust?'[32]

Behind the scenes at Hardie concern was mounting. Management worried that its mild warning labels and the continuing adverse publicity about asbestos could affect sales.[33]

Soon after the 1978 AGM, Macfarlane had spotted a chance for the company to completely reshape itself. A firm of stockbrokers had found what they thought was an ideal corporate match for Hardie. The giant British publishing and paper conglomerate Reed International was in financial trouble and seeking to offload some of its overseas operations. Hardie was one of the few Australian companies big enough to swallow its local assets. Macfarlane was initially cautious, but the more he crunched the numbers the more attractive the deal seemed. The Australian subsidiary, Reed Consolidated Industries (RCI), had earnings which were roughly compatible; like Hardie, it was involved in building products, but without asbestos, and its other activities offered promising areas for Hardie to eclipse its one-product image of the past.[34] Within four months, Reid announced what was then the biggest ever takeover of a listed company in Australian corporate history. Hardie had doubled in size, in one deal massively diluting its dangerous dependence on asbestos.

By 1980 Macfarlane was ready to take the final step. He set about eliminating asbestos altogether from the company's building products. Hardie had first invented a way to do this for its fibro cement sheets in 1964, when it developed the cellulose-based Hardiflex. But it had held off because it relied upon the mineral for the company's curved products like pipes, which had been its most profitable line. Now the managing director pressed ahead with a replacement program for flat sheets. Ignoring the protests of the company's old hands who argued asbestos delivered a superior product, he authorised the Western Australian Welshpool factory to run tests. In 1981 a cellulose product free from asbestos was rushed onto the market. The new Hardiflex had problems, exacerbated by the dry Perth weather, and within hours some of the fibre cement boards began to warp and pop their nails. Still, Macfarlane judged the effort worth the risk and was confident the Hardie engineers would respond to the challenge. They succeeded and Macfarlane extended the roll-out of asbestos-free Hardiflex II to the eastern states.

During this phase-out of asbestos, Reid maintained a brave public face. In a series of press interviews Hardie's chairman scoffed at the idea that the company was fleeing an unsafe product. 'Mr Reid denies firmly that the RCI takeover can, in any way, be interpreted as an escape from the sensitive area of asbestos — now on trial as a health risk,' reported the Melbourne *Herald*: 'Rather than brush asbestos aside, Mr Reid says, Hardie is a buyer of asbestos and is not directly involved in mining ... 'In the way that we use it, no one has found a satisfactory alternative,' he says ... Mr Reid says a lot of the so-called problems with asbestos have not been adequately analysed and explained.'[35]

He took a similar line with a *National Times* reporter. Was the rising world controversy over the use of asbestos a reason for diversifying, he was asked? 'Definitely not,' Reid retorted. 'The controversy has its origins in the US and the UK and there is a very important distinction between their industry and ours. Asbestos textiles and lagging were the great problems — industries we don't have here.'[36]

There was indeed a difference between the Australian industry and others. Because Hardie had marketed its products so successfully, Australians had become the highest per capita users of asbestos in the world. And although Hardie was not involved in textile production, since 1935 it had manufactured asbestos insulation, which was even more dangerous. Within a few years, Australia would attain the grim distinction of having the highest rate of asbestos-related disease in the world, at least three times the US rate. The asbestos 'problem' was, in fact, greater in Australia than anywhere else.

Later, Reid would admit the real reason for the corporate restructure: '... although we weren't saying it at the time ... if we hadn't discovered how to make asbestos-free technology, I'm not certain the company would have survived'.[37]

Hardie had already dropped the word 'asbestos' from the company name and from 'Asbestos House', its corporate headquarters. The next step was to pump millions of dollars into an advertising campaign to erase Hardie's connection to asbestos from public memory.

The advertising agency Coudrey Dailey, was briefed '... to distance the company from the adverse associations with asbestos'. It devised the slogan 'James Hardie — The Name Behind the Names', which was hammered unmercifully in the media during the next few years.[38] Hardie was projected as a patriotic, responsible and diverse company. It published books. It produced TV cartoons. It provided furnishings and wallpaper. It made pipes to water lawns and irrigate farms. The company's media exposure included sponsorship of *This Fabulous Century* and *Australians*, popular nostalgic series of Australiana by the former TV journalist Peter Luck, and the *Michael Parkinson in Australia* TV series, which was expected to have '... a high level of penetration amongst people over 18 in occupational groups which cover middle and senior management and professional people'.[39]

The advertising and sponsorship soon paid off. A survey commissioned by the company in 1980 revealed that 78 per cent of respondents had negative reactions to asbestos and associated the mineral with James Hardie. By the end of 1981 another survey found that the association between Hardie and asbestos had nearly halved to 44 per cent.[40]

Once the Hardiflex II asbestos-free sheets were rolled out into the major markets on the eastern seaboard the advertising strategy became bolder. In 1983, Hardie became the prime sponsor of the national fitness campaign, Life — Be In It. The irony could hardly have been more profound. As Hardie's in-house PR manager observed, this campaign '... has the advantage of being health and community minded and will offset any perceived undesirable aspects of our operations'. At the launch the campaign organisers claimed that James Hardie believed in a 'fitter, healthier Australia'. By the end of the year the board was informed that 'virtually all attitude measurements indicate ... a declining association with asbestos'.[41]

Behind this barrage of publicity, Hardie was still manufacturing asbestos products and continuing to lobby against the introduction of safer health standards. It successfully delayed a national ban on brown asbestos. Although the company announced in 1983 that it had eliminated asbestos from its building products, this was not entirely true. Its chemists had not yet managed to find a suitable alternative for Hardie's curved products like pipes. Corrugated asbestos cement for roofs and fences continued to be made in Hardie's South Australian factory at Largs Bay until 1984 and in Western Australia's Welshpool factory until 1986, while asbestos pipe manufacture continued in several states until 1987.[42]

Despite this ongoing production the company became sufficiently confident about its image makeover to publicly concede that its asbestos products may have been dangerous after all. Under the headline GOODBYE FIBRO, WELCOME PINO, the *Sydney Morning Herald* quoted a company spokesman as saying that although asbestos cement had generally been considered safe, '... it had been decided to replace all asbestos products to reduce the risk faced by tradesmen who often have to cut fibro using high-speed saws in confined areas'.[43] This was the risk that only four years before Reid had described as 'remote'.

Amid such spin, at least one reporter was persuaded that Hardie's involvement with asbestos had been minimal. Writing in the *Business Review Weekly*, Ross Greenwood announced that James Hardie Industries '... has worked solidly for years to overcome the stigma of being associated with the dangerous fibre ... [in the past] the health risks now identified with asbestos were not understood and generally accepted'.[44] Remarkably, Greenwood asserted that '... contrary to most people's beliefs James Hardie was never predominantly an asbestos company', an observation that contrasted starkly with Reid's description four years before of Hardie as a 'one product company' with a 'high dependence on one mineral'.[45]

At the same time as Hardie was diverting attention from its asbestos past, Reid also took steps to repackage his own image, hiring Eric White's director Les Anderson as his personal PR adviser. Anderson had left the PR agency to start a consultancy of his own, where he was joined by Frew, who continued his media work for Hardie while Anderson provided a specialised service for the chairman.[46]

Although Reid claimed to prefer to live in 'dignified obscurity',[47] on those occasions when he spoke publicly he was no stranger to controversy. 'Am I a conservative? I suppose that's fair comment,' he said to one journalist.[48] He had even weighed in on one of the biggest scandals to envelop the Australian business world during the late 1970s, when many Australian companies had avoided paying hundreds of millions of dollars of tax through so-called bottom-of-the-harbour schemes. A royal commission found the tax schemes, mainly driven by Western Australian-based accountancy firms, bordered on fraudulent. At a Congress of the Institute of

Chartered Accountants in Perth, the Fraser government's treasurer, John Howard, urged accountants to be more mindful of community expectations and not indulge in such blatant tax avoidance. But to reported cheers of delight from the audience, Reid departed from his prepared speech to declare that an accountant's duty was to the person paying the bill and government should fix the laws rather than 'lumber' one branch of the community with the task.[49] Later, the Labor opposition leader Bill Hayden singled out the Hardie chairman in Parliament for his involvement in a dubious tax-minimisation scheme, a charge Reid indignantly denied.[50]

As Reid set about recasting his image in a more benevolent light, some of the media coverage accorded to him bordered on fawning. He was described variously in newspaper articles as 'very much the gentleman's millionaire', and 'slight, always suntanned, always immaculately dressed'.[51] Despite his generally dim view of the media, he later advised aspiring directors: '... there are professional advisers, frequently ex-media contributors, who are very valuable in providing understanding of the way the media industry responds.'[52] The Hardie chairman evidently had found his personal PR coaching helpful.

In 1981, Reid turned his attention to lung diseases, donating funds and becoming inaugural chairman of the Institute for Respiratory Medicine at Sydney's Royal Prince Alfred Hospital. A few years later he also arranged for Hardie to finance research into mesothelioma at Western Australia's Sir Charles Gairdner Hospital.

Reid established a profile as a behind-the-scenes philanthropist and patron of the arts, a process made easier by the considerable fortune his family had amassed from asbestos. He became known as a generous donor to the Salvation Army, the Girl Guides, Opera Australia, the Art Galley of NSW, the National Gallery of Victoria, the Museum of Contemporary Art in Sydney and the Powerhouse Museum, to name just a few. The family trusts established by his father and uncle, as well as James Hardie Industries, are often included next to his name on the list of donors displayed on the walls of these institutions.

Not all of the chairman's philanthropy was 'obscure'. Unlike most of those from his uncle or father, Reid's donations had a habit of attracting naming rights. Thus at the Pymble Ladies' College, the John B. Reid building wing commemorates his role on the school's council. Across Sydney at the University of NSW, the John B. Reid Theatre marks his financial support for the Australian Graduate School of Management, where he became Emeritus Professor. The John B. Reid Fellowship also acknowledges his contribution to the Great Barrier Reef Trust. At Charles Sturt University students are awarded the Dr John B. Reid Scholarship.

His patronage of cultural bodies was recognised by Prime Minister Fraser in 1979 when he was appointed chairman of the Australian Bicentennial Authority (ABA), with an Order of Australia for 'services to industry' to follow the next year. Early into his ABA appointment, he opined to one journalist: '... despite their travelling and world exposure, Australians in many ways are not a civilised people. We are undemonstrative, inarticulate, and in some ways not a very powerful thinking people either.'[53]

The election of a Labor government led by Bob Hawke saw Hardie embroiled in renewed scandal over its former asbestos mine at Baryulgil, with Hawke bowing to Aboriginal pressure for a parliamentary inquiry into the conditions there. Aboriginal leaders called for Reid's sacking from the ABA.

Once again, Hardie's PR machine swung into action. The company commissioned its advertising agency Young & Rubicam Coudrey to counter the bad publicity expected when the inquiry concluded. The agency recommended 'positive' advertising to build on the 'Name Behind the Names' campaign: 'It is important *not* to be seen to be responding to criticisms arising out of the Baryulgil inquiry ... it would be preferable to commence the campaign prior to the findings and justifiably adopt a 'business as usual' stance.'[54]

The agency identified the 'managerial, professional sector of the community' as the group most likely to have 'the most immediate and perhaps most profound reaction towards James Hardie from the Baryulgil findings'. One way to create an 'attitude bulwark', it suggested, might be to sponsor an Australian art series, which could then be spun out into merchandising and competitions before being donated to charity. Hardie already held just such an art collection. With considerable fanfare, eighteen months after the parliamentary committee reported, Hardie announced a competition to find a suitable home for the James Hardie Library of Australian Fine Art, which would be donated to the nation. The company ultimately donated the collection as a bicentennial gift to the State Library of Queensland, where it was housed in specially constructed quarters which were later named the John B. Reid Room.

Criticism of Reid's position at the ABA continued. The following year, Reid waded into the education debate as chairman of the exclusive Pymble Ladies' College, describing the state school system as an 'intellectual desert'. It was too much for some Labor backbenchers, who joined the chorus demanding that he be sacked.[55]

The ABA became a highly controversial body. Its CEO, David Armstrong, had acquired a public reputation for profligacy, largely through his frequent overseas travel. In his absence, Reid began an affair with Armstrong's wife, Virginia Henderson.

Armstrong's numerous trips soon became a political liability. Prime Minister Hawke responded to public pressure by demanding Armstrong's resignation. Initially, Reid defended his lover's husband, declaring that he was happy for Armstrong to remain in his post. But within weeks Reid had sacked him on the understanding with Hawke that his severance pay should err on the side of generosity. It was indeed generous. Armstrong received a $500,000 payout under an arrangement whereby he only paid an estimated $7000 in tax.

When the details of the payout emerged in the media the political scandal became too hot for Hawke. A grim-faced Reid was summonsed to Canberra and, after a tense four-hour meeting with the prime minister, he too was sacked, despite an expression of full confidence in its chairman from the ABA's board.[56] One director in particular, Alan McGregor, who was married to Skye, Reid's cousin, was vocal in his defence. Soon after, Reid divorced Patricia, his wife of thirty years, and married Henderson. David Armstrong remained bitter over the experience, having lost both his wife and his job.[57] No details of this personal scandal were revealed by the media at the time.

Reid's political allies rallied around him. Within a year the conservative Queensland premier Sir Joh Bjelke-Petersen had appointed him to the board of Expo 88, where he served alongside its chairman, the former Queensland treasurer Sir Llew Edwards, in planning a Brisbane-based bicentennial exposition. Reid was soon to return the favour to his supporters, inviting first McGregor then Edwards to join the board of James Hardie.

Hardie's diversification had continued throughout the eighties with a string of other smaller acquisitions. At various times the company became Australia's largest book publisher, the country's largest manufacturer of envelopes, a maker of bathroom fittings and even the owner of Sydney's theme park, Australia's Wonderland. Some investments proved disastrous, others simply confused the portfolio. In the blur of corporate activity, though, Macfarlane clung to a vision that the company could revert to its core and develop the building products it knew best, now without asbestos. By 1984, the managing director was predicting an expansion into the US: 'We now have a board with which to tackle that market,' he enthused.[58] Five years later Hardie began constructing a fibre cement factory at Fontana in California, soon followed by a second plant in Florida, both producing Hardie's patented cellulose alternative to asbestos cement.

The Hardie board continued to change as John Reid shaped it to suit his vision for the company.[59] By the end of the decade asbestos had been eliminated from the company's products with scarcely an upset, unlike Hardie's bigger overseas partner Johns-Manville, which had sought Chapter 11 bankruptcy protection. The chairman told his shareholders with evident satisfaction: 'There is no doubt whatsoever if the transformation had not been handled intelligently and with great care, the damage to the company could have been very serious indeed.'[60]

Reid had continued to be involved in Hardie's asbestos negotiations during the transition, despite his other personal and professional distractions. Hardie had defused pressure from the unions in the late 1970s by agreeing to pay additional cash of $14,000 or more to sick workers on top of whatever they might obtain through the courts.[61] But in 1985, the Perth-based Asbestos Diseases Society of Australia (ADS) discovered that the money had been offered to an ex-employee on the condition that he signed a deed waiving any further right to sue. The Society went public, attacking the deed, and Reid twice called the ADS president to soothe his concerns.

'It's all been fixed with the union,' Robert Vojakovic recalls Reid telling him.[62]

The Hardie deed included strict confidentiality provisions designed to keep the practice secret.[63]

Secrecy, indeed, was Hardie's byword, one endorsed by the chairman, who would later advise aspiring directors to 'remain silent where there is criticism'.[64] Criticism of James Hardie could only get louder as the number of asbestos victims rose; Reid, however, saw the conflict more in class terms: '… the political philosophy of some people in the media see the creation of wealth as being immoral and greedy.'[65] Hardie did indeed create wealth for its shareholders, but it was at the cost of the lives of many employees and customers.

The Hardie chairman later proudly recalled a moment that led to one of the greatest spurts of shareholder wealth in the company's history, which occurred at the same board meeting at which he introduced its first female director, Meredith Hellicar, whom Reid had headhunted. A recommendation had been made to abandon Hardie's US operations, which by 1992 were making little headway against entrenched timber and mortar building interests. Reid and his in-law Alan McGregor 'protested vigorously'; within a few months the managing director was replaced with a new CEO, Keith Barton, who reinvigorated the US operations. Reid wrote in 2002: '... when 85 per cent of the company's business is in the USA and is very profitable, the wisdom of that decision ... is clear.'[66]

The strategy charted by Barton with his chairman's support also entailed the gutting of the asbestos-manufacturing giant, James Hardie & Coy. Now that the group had a market-winning alternative to asbestos they no longer had any use for the former asbestos subsidiary, which was increasingly the target of compensation claims from the dead and dying. Plaintiff lawyers like Armando Gardiman were later to describe this process as deliberate asset-stripping. 'If you look at the way they did it all — the sale of the technology, the payment of those enormous management fees and dividends, just enormous. At the end of the day they were left with shells that owned some real estate. And they sold the real estate, transferred that as well. There was nothing left. They got rid of the lot.'[67]

One by one, almost all of the subsidiary's assets, including its market-winning patents, were sold to other companies in the group, with the benefits ultimately accruing to the parent company, James Hardie Industries. A steady flow of dividends reduced the manufacturing giant to a shadow of its former self. Barton later asserted that each transaction was legitimate, and that 'there was never any suggestion that we wouldn't be able to meet our claims',[68] but two decades later, that was in fact what would happen.

As part of Australia's bicentenary, Reid commissioned a history of Hardie which was based largely on the research of a sycophantic writer, John Balmforth, who spent many hours interviewing former employees about their experiences. The factory workers did not have 'articulate or disciplined minds', Balmforth wrote to the company PR officer in charge of the history, but they nonetheless did have some views about asbestos and its possible dangers. When Hardie's lawyers suggested that his interviews '… might provide information of value to the company in defending any future actions which might arise and that possibly I could direct my questioning to that end', Balmforth confessed to some moral anxiety.[69]

Hardie's centenary book, Balmforth urged, should confront the asbestos health issue head-on:

> If the asbestos issue is ignored or essentially glossed over, the
> company's credibility on the matter would be prejudiced.
> Something is hidden only when there is something to hide …
> I'm not suggesting the company opens up a can of worms or
> debates the minutiae of whether a Baryulgil worker died of
> an asbestos-related disease or fell off the back of a truck …
> but it should read [as], and be, an open statement of the
> company's position and the world's state of knowledge about
> it down the years.[70]

A few weeks later Balmforth was replaced as the writer, and he demanded that Reid ensure that his name would in no way be associated with the book.[71] When the book was published, almost the only reference to the dangers asbestos made in the 252-page history of the Hardie empire is found in Reid's foreword: 'The book deals only briefly with asbestos. The company made products that contained asbestos, and as a result is involved in legal proceedings. We therefore believe it would not have been proper to discuss these matters, however seemingly unrelated.'[72]

And what became of the vast number of files on asbestos that had accumulated, not only from Balmforth's research, but from the company's official records? Hardie claimed that it kept all these files to assist its defence against compensation claims. Some were destroyed during floods in its Parramatta repository, and the majority were eventually transferred to Sydney's central business district where they have continued to be provided as evidence in litigation.

But not all of Hardie's relevant asbestos documents have been declared to the courts. Many, such as Balmforth's interviews and the files of the PR agency Eric White, were donated to the State Library of New South Wales as part of the vast Reid collection, buried among tonnes of uncatalogued records about the family and company history. Although Reid availed himself of the tax concession allowed for such a gift, he declined an offer of a public ceremony. 'I don't think a public handover today would do much for the State Library or James Hardie,' he wrote.[73]

In 1999, teams of lawyers for Hardie combed through these files, in the state library, searching for items of use in defending against asbestos claims. Two photocopying machines were hired to assist the mammoth task. Eventually the company was given permission to remove boxes overnight to speed the process: nearly a third of the collection was removed and returned.[74] Despite this effort, significant documents in the library's Reid collection were still not declared by the company during subsequent discovery proceedings by asbestos claimants. When quizzed in court the counsel representing Hardie's former asbestos company put the omission down to 'stupidity', 'laziness' or a 'legal misjudgement'.[75]

There are still some important documents, like Neil Gilbert's warning to the company to abandon asbestos, which have never emerged.

In 1996, after nearly a quarter of a century at the helm of the company, John Reid retired. He had led Hardie through its greatest asbestos expansion, despite warnings at the time of the dangers. He had then overseen its transformation away from the toxic mineral, while the company crafted a public profile and legal strategy that kept the death toll from its products as secret as possible. When the legal tide began to turn, it was under his chairmanship that the company began to quarantine its asbestos subsidiaries as it took its asbestos-free product into the US market. He'd chosen the chairman, CEO and board of directors to continue the process. Now was time to reap his rewards in retirement as the company's largest shareholder, occasionally providing inspiration for corporate leaders with his fireside chats at the residential schools of the Institute of Company Directors. He had never answered the questions about asbestos that needed to be asked; questions such as why the company had refused to phase out its use earlier. With any luck, he might never have to.

eight

A RACE AGAINST TIME

When 300 unionists in 1989 banged on the large glass windows surrounding the 'fishbowl' meeting room at Vales Point Power Station in the Hunter Valley, the visiting NSW energy minister, Neil Pickard, trapped inside, must have wondered what had hit him. The workers had surrounded his official party, blocking the doors and corridors. According to one of the demonstrators, Allen Drew, 'We wouldn't let them out. Or rather, no one was game to come out!'[1]

The workers were outraged by the spectacle of their workmates with mesothelioma struggling to live long enough to secure compensation for their families.

'We took the opportunity to go off the job at the same time as the Minister came,' said Drew, a delegate for the FEDFA, the union representing the workers who wrapped the asbestos insulation, or 'lagging', around the hot pipes to the station's turbines. The union was demanding the government find a way to speed up the compensation claims of workmates with asbestos-related illnesses.

Those laggers who had contracted mesothelioma faced a race against time to win compensation for their families before they died. The judiciary was slow to react. One NSW Supreme Court judge had even reserved his decision for more than twelve months and the claimant died before it arrived. The sight of pain-stricken asbestos litigants struggling for breath while their case was heard became more frequent. Not all could marshal the strength to stay alive for the duration of their case.

Within months Premier Nick Greiner's Liberal government had approved and set up a new specialist court, the Dust Diseases Tribunal, to fast-track such cases. The tribunal was the first court of its kind in the world to develop special procedures to suit the unique nature of asbestos diseases.[2]

The laggers were soon to be joined in the courts by thousands of plumbers, carpenters, builders and home renovators who would die from the fatal dust they had inhaled from Hardie products, despite the company's repeated assurances that this could never happen.

Before the tribunal was created, it was a battle just to get into court, as the lagger Ronald Baker had discovered. Baker had left school for a job at Vales Point Power Station and like most laggers he worked for several of the specialist firms installing asbestos insulation. He was forty-one when he developed mesothelioma. As the countdown to his death began, his lawyer, Turner Freeman's Armando Gardiman, began a frantic search to trace his former employers. Two of the firms were defunct; the third had returned to England. Only one had traceable insurance, but even that was for a paltry amount. Luckily, a workmate remembered that Hardie had supplied the insulation and so the company was cited as a defendant.[3]

Because Baker's fatal exposure had occurred many decades before, the statute of limitations barred him from launching an action unless a judge granted an exemption. In court, Hardie opposed Baker's extension of time. The company's senior counsel suggested that Hardie could never have known about asbestos dangers when Baker was employed. Baker's medical expert[4] disagreed, remembering that he had examined a group of men in the late 1950s suffering from asbestosis, although he assumed confidentiality would prevent him from naming the firm involved. The judge assured him that he could. It was James Hardie.

Although Baker won the right to pursue his claim in the courts, it was to no avail. Hardie appealed, and while the company lost, the delay proved critical. Baker died. Gardiman refused to give up, relaunching the action on behalf of Baker's wife. The case dragged through five court hearings, including the High Court, in a bitter legal battle. After four years Baker's wife and his two children were compensated for his premature death.

There were many more cases to follow. What began as a trickle of sick laggers turned into a steady stream. After Gardiman realised Hardie had been Australia's sole manufacturer of pre-formed insulation he began to sue the company as a co-defendant in every action.

When the Dust Diseases Tribunal was set up neither the Liberal government nor the tribunal's first president, John O'Meally, had expected it to last more than a few years. O'Meally, who had no experience of asbestos cases other than a brief stint as counsel for Hardie before joining the Bench, was on a steep learning curve. It quickly became obvious that the state was in the midst of a growing epidemic of asbestos diseases. The tribunal hearings revealed in stark relief the distinctive suffering such diseases caused, shocking even some of the case-hardened lawyers involved.

Said one: 'What really got to me in the early days was the way that people were dying in utter agony, seemingly in uncontrolled pain. And many of them too young to die.'[5]

To give the breathless claimants some relief a sick room complete with its own oxygen supply was set up in a room adjoining the court. O'Meally soon went one step further and took the proceedings to the patient, establishing bedside hearings like the one held for Bernie Banton. These hearings were often tragic scenes. As one participant observed:

> I remember once going to a caravan park and there was an old fellow there dying, and I think there were ten lawyers, ten defendants, each cross-examining him. He died that night, the poor fellow, and he was moribund. But the obscene part about it was that he was living in this caravan park, at the gates of which were two Mercedes, a Bentley, a Porsche ... all the lawyers.

O'Meally made every effort to finish cases before the plaintiffs died. Speed was of the essence. The ever-increasing flood of claims led to an unprecedented schedule. On occasions, the court would convene at seven o'clock in the morning and run until four in the afternoon. After an hour's break the tribunal would reconvene for a second case that might run until ten o'clock at night. Hearings ran six days a week, even on public holidays. A court official said that the plaintiffs, though mortally ill, appeared determined to seek justice.

'It's pretty interesting to see the way people hang on, you know. They can keep themselves alive. And then they go once the case is finished.'

To speed up the process and underline his support for the new tribunal, Attorney-General John Dowd declared a three-year amnesty to the statute of limitations for asbestos claimants. Within a few years the number of cases listed before the tribunal snowballed. A new judge, Peter Johns, who, like O'Meally, had previously worked as a Hardie counsel, was appointed to assist handle the workload.

An amendment was also made to the tribunal's Act which meant that issues already argued out in previous cases did not have to be argued anew. This change was triggered by a dramatic claim in 1994 against CSR by Vivien Olsen, a thirty-four-year-old who was born in Wittenoom. Olsen, then one of the youngest people to contract mesothelioma in Australia, had worked as a research officer for the former NSW Labor Attorney-General Frank Walker. Her first few years of life in Wittenoom, in an environment heavy with blue asbestos dust, proved fatal. Her mother would lie her on a bunny rug in a garden lined with asbestos tailings when she hung out the washing and a thick blue ring stuck to the edges of the tub after the children's bath.[6]

Olsen's legal proceedings were agonisingly protracted. If she died before judgement the vast bulk of her damages died with her. As her condition steadily deteriorated, the tribunal spent three weeks taking evidence it had heard many times before: evidence about when and what CSR had known about the dangers of asbestos. Judge O'Meally sat on Christmas Eve to ensure that a verdict was entered before she died. It was a Saturday: remarkably, O'Meally delivered a verbal judgement, occasionally referring to notes, from nine o'clock in the morning until three o'clock that afternoon. Olsen died a fortnight later, O'Meally's judgement survived a subsequent appeal.

* * *

The impact of asbestos was now becoming notorious in the Australian media and elsewhere as it claimed an increasing number of lives, not all confined to mines or factory floors. Shock waves rippled throughout New South Wales when in May 1990 Premier Nick Greiner tearfully announced that a man widely admired by both sides of politics, the state's governor, Sir David Martin, had succumbed to mesothelioma. Martin had asked for his condition to be publicised to alert others to the hazards of asbestos. His fatal asbestos exposure was believed to come from his service in the Royal Australian Navy, where asbestos insulation (some supplied by Hardie) was wrapped around the steam pipes on board ships.

Not all journalists were in favour of the tribunal. A conservative commentator from the *Australian Financial Review*, Paddy McGuinness, articulated the alarm felt within Hardie about the new court. He reported complaints that decisions of the Dust Diseases Tribunal 'reflect bias in favour of complainants', in an article that took aim at 'captive' state tribunals and the appointment of '... increasingly political and ideologically committed judges, many of whom are really political activists'.[7] In a subsequent column McGuinness sneered at the developing 'milch-cow' of product-liability litigation, arguing that it was 'hardly sensible' for asbestos sufferers to obtain damages from Wittenoom where, he claimed, production had only continued because of the demands of the unions.[8]

Hardie's lawyers, nevertheless, adapted their strategy to deal with the changing political and legal environment. During earlier times the company had encouraged secret deals with those who fell sick even before they had consulted a lawyer.[9] Hardie had also referred its ailing staff in NSW to the no-fault Dust Diseases Board, where compensation was cheap and there were no public proceedings. That, generally, was the last the company heard of the matter.

In the days before the tribunal, those who sued Hardie for negligence in open court proceedings were vigorously opposed, something the Baryulgil people learnt to their cost. Hardie took advantage of the delays in the court system, as David Say, a former managing director, recounted: 'It was a horrible situation. The award to someone who has died (or their relatives) was always lower than if they were still alive, even if they had hours to live. The obvious temptation for lawyers from the insurers was to spin it out.'[10]

The company knew what to expect once disease spread beyond its factories. It had seen an avalanche of litigation begin to engulf its partners in the US and it prepared for a similar onslaught in Australia. Hardie locked in former executives who might give evidence against it. Neil Gilbert, the former engineer who had urged the company to abandon asbestos before he left in 1970, was offered a year's salary to come back to 'look after the asbestos problem'.[11] He refused, but later learnt that several of his former colleagues, Dr McCullagh amongst them, had accepted annual consultancies. Gilbert had no doubt why they were offered. 'That was a bribe to keep them quiet. I believe they were doing everything they could to keep the lid on it,' he told me.

'Keeping the lid on it' became a priority for the Hardie lawyers facing the tribunal. They rarely allowed a judgement to be recorded against the company. So long as Hardie was not the only defendant in a case, its lawyers would encourage confidential settlements with claimants. The company would invariably contribute to the damages claimed rather than see a case run to a conclusion. This willingness to settle cases also suited the major plaintiff law firms specialising in asbestos litigation. As Turner Freeman's Gardiman noted: 'Most of the cases settled without a shot being fired. Some ran for a day, or part of a day, until the contribution was sorted out. But they never went to judgement.'[12]

Hardie had good reason to fear court scrutiny. As the largest asbestos manufacturer it was likely to become the most frequent defendant. The more cases which went to judgement, the more publicity was likely, which would encourage still further claims and increase the prospects of damaging evidence emerging against the company.

Gardiman discovered another reason why Hardie was so eager to settle. During a case in 1990 involving a carpenter who had developed mesothelioma after using Hardie's 'Super Six' corrugated asbestos cement sheets, Gardiman suggested to Hardie's lawyers that they should cross-claim against CSR. They expressed no interest, he recalled, saying they preferred merely to settle the cases with contributions from other defendants.[13] The reason why dawned on him later.

Almost all of Hardie's asbestos cement products manufactured between 1957 and 1968 contained a percentage of the same Wittenoom blue asbestos that had killed Vivien Olsen, which Hardie had been pressured to buy from CSR under the threat of a tariff on its imported fibre. Wunderlich, Hardie's smaller competitor, had not used blue asbestos. For a decade of massive building growth when Hardie's asbestos cement was widely used, the company was the only manufacturer using the type of asbestos most likely to kill. Said Gardiman: 'Fibro homes and factories were being built by their thousands all through that period. And if suddenly all these people ... were coming down with mesotheliomas, there was going to be one obvious target.'[14]

The plaintiff lawyer was content simply to file this knowledge away in the back of his mind. He had developed a good working relationship with the solicitors handling the Hardie cases, and while the company kept settling there was no reason to make its life any more difficult. He had no wish to see a repeat of the US experience, where thousands of claimants had been left with little or no compensation after asbestos giants like Johns-Manville obtained bankruptcy protection.

But when the Hardie board decided on a switch in legal tactics in the late 1990s the cordial relationship between Hardie's lawyers and the plaintiff firms evaporated. Spurred by the rising costs of asbestos claims, it chose a tougher approach to 'understand the extent of its liability', according to Meredith Hellicar.[15] At the suggestion of one of Australia's most prestigious law firms, Allen Allen & Hemsley ('Allens'),[16] Hardie launched a series of test cases. Gardiman later described it as an 'absolutely disastrous' decision for the company, but both he and his colleagues were soon to feel the blowtorch of a hardline strategy wielded by one of the country's biggest law firms. Turner Freeman installed a new fax machine to handle the extra paperwork. The struggle became very personal:

Allens advised them to fight. They just wanted to litigate everything. Everything became an argument. You had to produce the last bit of paper to get it across the line. I'm sure they thought, 'If we absolutely swamp them in paper, they'll give it away.' I think they vastly underestimated the plaintiff lawyers, because the people who worked here were really committed. People like Joe Calabrese — his father had severe asbestosis. I mean, he had a personal interest in getting these people.[17]

The Hardie lawyer then in charge of its asbestos litigation, Mark Knight, described the new strategy rather differently. For him, it was a reaction to 'the greed factor' of plaintiff lawyers.[18] Knight, a prickly man, was driven to talk to me years later by a strong sense of indignation. He looked back on the period with ill-concealed bitterness.

'I don't care what anyone else says, James Hardie did business honourably and tried to be as dignified as the circumstances permitted for these cases. Until the point came when our good nature was either used ... or perhaps a better way of putting it was that they just simply wanted to go too far, and I don't understand why.'

'Honourable' was scarcely the term used in the offices of Turner Freeman to describe Hardie's actions. There, the impact of asbestos disease was felt on an almost daily basis as victims continued to arrive on the doorstep. The firm's biggest client, the Australian manufacturing workers' union (AMWU), had just witnessed the death of its national president, Brian Fraser, from mesothelioma. A giant of a man, Fraser had shrunk to a few stone as he clung to life in his hospital bed while Gardiman negotiated a settlement for his family.

The first battle in the tribunal proved an unmitigated disaster for Hardie. John McCusker, a carpenter with mesothelioma, had used asbestos cement products made by both Hardie and its competitor, Wunderlich. On his behalf Gardiman claimed compensation from both firms. Wunderlich offered to settle for 15 per cent of the damages and costs, but Hardie was not interested. Its new lawyers at Allens wanted to run with the case.

Gardiman had an ace up his sleeve: the knowledge that Hardie was the only asbestos cement company that had used the deadly Wittenoom blue. He presented the court with the shipping consignment notes from Wittenoom, revealing that tens of thousands of bags of its blue asbestos had arrived at the Sydney wharf destined for Camellia, where it was mixed into the Hardie product.

'The whole theory was to just run blue at them. We just kept introducing all this evidence about the blue, introducing documents about the blue, and the amount of blue … it was just blue, blue, blue.'

It soon became a template for future mesothelioma cases at Turner Freeman:

> It meant that if you had any carpenter come through the door, the first question you asked him was: 'Now did you ever work with Super Six? What about Log Cabin? Hardiplank?' And you worked your way through the moulded product. All of the corrugated and moulded product had to be cut with a power saw. And the moment you had power saws, you had clouds of dust …

Weeks into McCusker's case a Hardie witness, the former Victorian health department officer Janet Sowden, made a crucial admission under cross-examination. Was it foreseeable by the mid-sixties, she was asked, that anyone cutting asbestos cement sheets containing blue asbestos with a power saw could be seriously injured?

'I think anybody with a modicum of knowledge would've thought that, yes, power sawing with anything containing crocidolite [blue asbestos] would have been potentially harmful,' she replied.[19]

'The moment she gave that answer they were gone,' said Gardiman. 'Wunderlich pulled their money out and said [to Hardie], 'We're out of here!''

Looking back, Hardie's lawyer Mark Knight could only snarl with frustration. The company had never wanted the Wittenoom asbestos in the first place.

> The federal government said to James Hardie: you will buy this stuff [blue asbestos] even though it's crap ... even though it's not the right fibre for your products, you will buy it or we will slap a big tariff on you. So why doesn't CSR get dragged into these cases more? They encouraged and led the federal government into this decision. And they're the ones who are still specifying blue asbestos in the 1960s for their wretched wheels on their trains. Beats me![20]

Turner Freeman rarely sued Wunderlich again. There was no point. All they did was sue Hardie and leave it to the company to claim money back from other producers. As Australia's largest asbestos producer, Hardie had always been likely to become the country's biggest asbestos defendant: McCusker's case now ensured it.

Chairman John Reid's words in his 1978 annual report

lingered with a menacing resonance: 'Every time you walk into an office building, a home, a factory; every time you put your foot on the brake, ride in a train, see a bulldozer at work ... the chances are that a product from the James Hardie group of companies has a part in it.'[21]

What made it worse for Hardie was that, for the first time, hundreds of its confidential documents discovered during the trial were now on the public record at the tribunal, available for use in any subsequent proceedings.

Years later, Hardie's asbestos litigation lawyer, Wayne Attrill, pondered over the company's deteriorating claims profile. Why was Hardie increasingly being sued as the sole defendant? he wondered. Why weren't Turner Freeman suing Wunderlich anymore? He put the questions directly to Gardiman at a city coffee shop. Gardiman explained that since the evidence had gone on the record that Hardie was the only manufacturer to have used blue asbestos he could not see any point in suing anyone else.

'They were in fact the victims of their own litigation strategy and they now just had to wear it and accept it. I said to him, 'You can't turn this back, you can't rewrite the facts. They speak for themselves.'[22]

Following the end of Dowd's amnesty on the statute of limitations, claimants were once again struggling to have their cases heard before they died. The tribunal's president, John O'Meally, recommended further changes to the law in 1994, which were accepted by the Liberal government, but the legislation stalled in the Cabinet Office. An election intervened, replacing Premier John Fahey with Labor's Bob Carr.

The new attorney-general Jeff Shaw changed the tribunal's rules to enable 'provisional' awards, which allowed claimants to return for additional damages if they contracted a subsequent disease, such as cancer. Bernie Banton would become the first asbestos victim to test this provision in court years later.

But Shaw was soon to embrace a raft of further changes to the DDT. He had met with leaders of the AMWU following the death of its national president and attended the openng of a new office for the asbestos victims group, the Asbestos Diseases Society of NSW,[23] where he spoke to some of the mesothelioma sufferers and their families and was briefed on the need for reform by their lawyer.[24] Shaw agreed to abolish the statute of limitations for all dust disease claims and to ensure that general damages — the largest of the compensation payments — should survive a claimant's death. Both measures would lessen the need for the increasingly common deathbed hearings, but both were likely to increase Hardie's costs.

Initially, the attorney-general had difficulty getting the changes listed for consideration by Cabinet. Hardie had combined with CSR to lobby against the proposed legislation and complained that the industry had not been consulted. The two companies presented their case to Shaw, who was polite but unmoved.

Hardie hired the PR consultancy Hawker Britton to gather intelligence and assembled a coalition of companies with exposure to asbestos claims to help, including CSR, Orica, Comalco, AGL, Boral, South Pacific Tyres, Pioneer and the Insurance Council of Australia. It also prepared a thirty-four-page submission opposing the changes and circulated it to members of parliament.

After months of lobbying the campaign proved an expensive

failure. The asbestos victims group ADFA and the unions had countered with lobbying of their own and the attorney-general and government stood firm.

Hardie then pinned its hopes on upper house independents to block the changes. It was in for a rude shock. The crucial debate, the company's lawyer reported, was characterised by 'emotional and extravagant language' and 'wild accusations', with one cross-bencher suggesting that companies like Hardie 'should be on trial for murder' and another that its executives should be 'subject to criminal charges'.[25]

The independents voted unanimously in favour of the bill.

With a degree of understatement Hardie's lawyer advised his board: '… we will need to address the fact that decision-makers including the Government, parliamentarians and bureaucrats do not at present regard James Hardie as a credible source of information on the operation of the dust diseases compensation system in NSW.'[26]

The experience reinforced the conviction of Hardie's board that the sooner it could be rid of the company's asbestos past, the better.

Not all of Allens' test cases failed. One was to underpin the legal basis for Hardie's future corporate strategy, by reaffirming the corporate veil protection the parent company had acquired in 1939 when it moved its asbestos operations into subsidiaries.

The case involved Desmond Putt, who had worked for four years at Hardie's Penrose factory in Auckland, New Zealand. Fifty years later Putt had mesothelioma. He was unable to claim from Hardie's local subsidiary because in 1972 New Zealand's Labor government had established a no-fault compensation scheme which barred common law negligence actions. But Putt could try to sue the Australian parent company, James Hardie Industries Limited, and in 1997 he lodged a claim before the NSW Dust Diseases Tribunal (DDT).[27] If he were to win, Putt had to pierce the 'corporate veil'.

Winning was not an impossible task. There were precedents that involved Hardie's former competitor CSR. In 1988, Peter Gordon, from the large Melbourne-based plaintiff law firm, Slater & Gordon, had won a series of landmark victories on this very point against CSR on behalf of Wittenoom miners and residents. A fully owned CSR subsidiary ran the Wittenoom mine, but Gordon's legal team successfully established that not so much as a packet of paper clips could be ordered by the Wittenoom subsidiary without approval of CSR's head office in Sydney. Three verdicts were handed down awarding six-figure sums against the company. For one ex-miner, Klaus Rabenalt, the jury added $250,000 in exemplary damages as punishment for what it found to be CSR's 'contumelious disregard' for his safety. Within a year, CSR had abandoned the corporate veil defence and negotiated a global settlement with Gordon for more than 200 claimants.[28]

In NSW, the DDT's Judge O'Meally had also upheld a claim against CSR by Sydney factory worker Norman Wren, who had developed asbestos disease after working in another subsidiary company. Again, because CSR managers had been involved in day-to-day operations on the factory floor, the tribunal ruled they were directly negligent through their proximity to the operation.[29]

Putt's legal team managed to convince Judge O'Meally that a similar situation prevailed in Hardie's New Zealand operation, but the Court of Appeal overturned the tribunal judgement.

To appeal to the High Court, Putt retained the senior NSW counsel David Jackson as his lawyer. Ironically, four years later, Jackson would loom large as the special commissioner investigating the scandal that enveloped Hardie's move offshore and the near bankruptcy of its asbestos foundation. Jacksons's efforts were unsuccessful, with the High Court refusing leave to appeal, although in its deliberations it hinted that other arguments against the Hardie parent company might have prevailed.[30]

This win in the highest court of the land underlay Hardie's future legal and political strategy. The company argued that regardless of claims against its former subsidiaries, the parent company was untouchable. Yet privately its lawyers had misgivings that echoed the hint given by the High Court. Many of the key Hardie personnel associated with managing asbestos exposure and disease, like Dr McCullagh, were employed by the parent company, James Hardie Industries. There was also a great overlapping of directors between the parent and its offspring companies during nearly half a century of asbestos production. A successful claim against James Hardie Industries was still possible.[31]

Belatedly, Hardie's top executives began to realise that using these hardline tactics before the tribunal may have been counterproductive. The company's legal position had continued to deteriorate. Increasingly Hardie was the sole defendant in asbestos claims. Halfway through 1999 it was paying roughly half the defence bill; by the end of the year it was three-quarters, and the average cost of its settlements had jumped by two-thirds.[31]

Hardie still had one other card in play. If it had failed in its duty to care for its employees or customers, its lawyers reasoned, so too had the New South Wales government. The company began gathering evidence for a test case that, if successful, would give the government a strong incentive to change what it called the 'plaintiff-friendly' compensation system. Mark Knight later summarised the argument:

> The state government knew that asbestos was dangerous in 1927. Why didn't they stop it? ... I don't understand how a state government can stand up today and say, 'Well, it's all James Hardie's fault ... ' Now why is that, when it was the state government and the federal government who were forcing James Hardie to use asbestos?[33]

Hardie, supported by its legal advisers at Allens, determined to test the issue — if necessary, taking it all the way to the High Court. The company's lawyers combed through the state archives to find evidence that the government had known of the risks of asbestos, but had done little to control its use. If Hardie could show that the state itself had failed in its duty to keep workers safe, it could lodge counterclaims in virtually every case. Most importantly, Hardie would have a potent argument to convince the government to wind back the tribunal.

And Allens thought it had just the right case to win. Warren Hay had worked for three years as a fitter's mate during the construction of the Wallerawong Power Station near Lithgow, NSW, where in 1958 he helped install coal-crushing mills under the station's giant steam boilers. Thirty metres above him laggers wrapped the boiler pipes with Hardie's K-Lite insulation, which contained 15 per cent brown asbestos. K-Lite was supplied with a company report asserting that the product could be 'cut, scored and sawed with the normal tools of the trade. It is non-irritating and non-toxic.'

Hay's story was depressingly familiar. As the laggers applied insulation, clouds of asbestos dust fell to the ground. He later told the tribunal: 'Fine asbestos particles could be seen everywhere ... sometimes the ground was white with particles which were so concentrated they looked like snowflakes.'[34]

Allens lost Hay's case in the tribunal. Judge Curtis observed that if Hay had sued the state it, too, would be liable, but Hardie had created the danger in the first place. He ruled that: 'James Hardie made large profits from selling vast quantities of asbestos heedless of the dangers to others ...' It was neither 'just nor equitable' if the loss from those activities should be borne '... not out of James Hardie's profits or risk capital, but by the taxpayers of New South Wales'.[35]

For Hardie's lawyers these comments merely confirmed the tribunal's 'underlying antipathy' to the company.[36] They lodged an appeal and vigorously pursued the case over the next four years, well aware of the gun they were pointing at the NSW government.

During the next few years both Turner Freeman and Slater & Gordon established interstate offices to satisfy the exploding demand for specialist asbestos compensation lawyers. Turner Freeman's partners, realising that many baby boomers exposed to asbestos as workers in New South Wales had moved to Queensland on retirement, opened a Brisbane office that soon began to attract a steady stream of sick clients. Invariably, they were former users of Hardie's asbestos cement.

* * *

Peter Russell, Hardie's former safety officer, was to provide critical evidence about Hardie's failure to warn its asbestos cement customers, which went onto the tribunal record for future cases. Russell appeared as a witness for Peter Thurbon, who had established a business building and renovating houses while his wife Elizabeth worked full time in Canberra.[37] The couple were hoping to save up a 'nest egg' to pay for their children's education and their own retirement. In 1999, their dream was shattered. Elizabeth recalled:

> One day he goes into hospital with a cough and they tell me he that he only has six months to live with a thing called mesothelioma. I'd never heard that word before! And you can't just grasp that your whole life, in one word, has been ripped to shreds. People are given six to twelve months to live and in that time they have to live their whole retirement. They have to pay off their mortgage, help their student children and live themselves while they've still got a little bit of breath to do it.[38]

Allens' conduct during the Thurbon case suggested that its lawyers had learnt little from their earlier failed defences. Initially Hardie set out to discredit Thurbon, hiring a private detective to prove he was an incompetent builder. Turner Freeman responded by calling Russell to give evidence. He had testified in two earlier cases, but this was the first time he had appeared before the tribunal. Hardie's counsel attacked Russell's credibility, insinuating that he had become a professional witness motivated by a grudge against the company, and to prove it, he cross-examined Russell on his previous statements. But Russell withstood the onslaught well: Hardie's lawyer later reported to his board that he had made 'a convincing witness'.[39]

What was more, his evidence was now on the tribunal's public record available for use in future cases.

For a tearful Elizabeth Thurbon, Russell's evidence had come as a revelation.

'He begged James Hardie to put warnings on their products. They wouldn't put in wet processing in their plants. They employed older men over fifty to sweep the dust so they'd die of something else before everyone knew they had asbestos cancer. But they wouldn't put a little label on their product.'[40]

Her tears were not from grief. They were tears of anger.

Allens' make or break tactics had a serious downside that Hardie had yet to fully appreciate. Not only had the failed test cases ripped away much of the secrecy that had shrouded Hardie's former asbestos operations, they had alienated both judges and politicians. By the end of the nineties, the company found itself increasingly isolated in a litigation landscape that would only get worse. Asbestos disease was spreading in the community, its victims were getting younger and their lawyers more experienced.

nine

PLOTTING FROM THE WAR ROOM

The year was 1998 and the board of James Hardie Industries Limited had given the go-ahead to a plan to relocate headquarters to the Netherlands, where it hoped to pay less tax on dividends it would receive from a new company to be formed to run its burgeoning US business.

A key element of the plan was to shed its asbestos liabilities by hiving off Hardie's two former asbestos subsidiaries — the now defunct brake-lining company Jsekarb ('brakes' spelt backwards), and the much larger manufacturer of fibro cement, James Hardie & Coy (called 'Coy' for short) — from their parent.

The James Hardie spinners began to meet secretly after work in the basement of the old Asbestos House, the company's Sydney headquarters. They would leave their offices casually at five o'clock as though they were going home and then quietly regroup downstairs in the 'war room'. The room was behind an unmarked door where nobody was likely to notice their frenetic activity.

Those in the PR team were the only ones in the company who knew of the relocation plan, other than the board and a handful of senior executives. If word leaked out, they believed the move would be doomed before it began because of the anticipated outrage from asbestos victims and their representatives. Their job was to craft a communications strategy to help it succeed. So sensitive was their task that Hardie's new general manager of corporate affairs, Greg Baxter, did not even mention asbestos in his brief for the board, where he referred to asbestos by the codename 'Apple':

> ... the separation of operating businesses from contingent liabilities could arouse suspicion, criticism and opposition ... Critics will have numerous public forums, such as a willing media, in which to air their concerns, particularly in relation to Apple and tax.[1]

Baxter, a tough, phlegmatic media-minder, was part of a new management team installed when Hardie began to move into the huge US market. Other key figures in the team included Peter Macdonald, then head of US operations — a reformed smoker who joined conference calls from his Californian home while on the exercise treadmill — and Hardie's general counsel, Peter Shafron, an extremely bright, affable lawyer recruited from Allens.

John Reid, who had supervised the company's transformation, had recently departed after nearly a quarter of a century as chairman, while still remaining one of Hardie's most significant shareholders. Following Reid's divorce settlement with his first wife Patricia, a large parcel of Hardie shares had been sold on the Australian Stock Exchange, enabling the corporate raider BIL, run by the New Zealand multimillionaire Sir Ronald Brierley, to secure a slightly larger shareholding than that of the Reid family.[2]

Nonetheless, Reid had anointed his friend and in-law Alan McGregor as the new Hardie chairman.

McGregor made no secret of his attitude to asbestos claimants. He thought the sooner the company shed its asbestos liabilities, the better. Not long after Reid invited him to join the board in 1989 he had said as much, according to Baxter's predecessor, Jim Kelso: 'He [McGregor] had the view a long time back that the success of the company would be if we could eliminate the need for expenditure on asbestos compensation. That was the opinion that he voiced to me then. And I presume that he held that all the time.'[3]

Although Hardie directors were later to stress that Hardie's growing asbestos liabilities were only a minor factor in their considerations, the board was increasingly alarmed by the recent experiences of the group's former asbestos companies before the NSW Dust Diseases Tribunal (DDT), where Hardie's damages bill was steadily rising. Its complaints about the tribunal had fallen upon deaf ears within governments of both persuasions.

Hardie's asbestos-free operations in the US were booming, generating more than three quarters of its profit. Yet in the company's Australian home base, asbestos was becoming a worsening problem. Not only were the cost and number of claims going up, Hardie's insurer QBE was refusing to pay for the ever-increasing negligence claims from those injured by its products.

QBE had taken fright following the first claims Hardie received from customers in the early 1980s, rescinding its policies and declaring that Hardie had failed to disclose knowledge about the dangers of asbestos. The insurer smelt a rat because in 1952 Hardie had taken out unlimited liability cover, rather than the minimum that its competitors took when such insurance became compulsory in New South Wales. QBE's alarm increased when Hardie received a claim from the US, where the company had exported a small quantity of asbestos cement and brake blocks during the 1960s and seventies.

Hardie sued QBE for cancelling its policies. In a preliminary judgement the NSW Supreme Court Justice Rogers warned that the case could involve 'very large amounts of money indeed', with the outcome 'of immense consequence to the future well being of the parties'.[4] Both companies immediately issued statements downplaying the impact. Hardie's spokesman Jim Kelso told one newspaper that 'claims for asbestosis in Australia bore little resemblance to those in the US ... where contact with asbestos was much more harmful',[5] repeating an earlier suggestion of Reid's that local exposure to asbestos was less serious than the overseas experience. Yet Australia was already well on the way to having the highest per capita incidence of asbestos-related disease in the world.

Nonetheless, the publicity caused the share price of both companies to plummet. The two firms quickly entered private talks, emerging two years later to announce an interim agreement, under which QBE would pay half of Hardie's claims to a limit of $4 million. That deal had now expired and since June 1995 the insurer had refused to pay anything more. After another year of talks QBE's best offer was an additional $5.8 million.[6]

Just as their predecessors had done, Baxter and the company's general counsel Peter Shafron prepared answers to any awkward questions about asbestos that their chairman McGregor might be asked at Hardie's forthcoming annual general meeting:

James Hardie deals with all claims in the most expeditious
and sympathetic manner possible. James Hardie settles the
overwhelming majority of claims out of court so as to
minimise unnecessary legal expenses, bedside hearings and
delays ... based on past experience, we do not believe
pending or future asbestos litigation will have a material
adverse impact on the Company's financial position.[7]

Behind the scenes, the company was not nearly so sanguine
about its future asbestos liability. In Baxter's communications
strategy for Project Chelsea — as the project to move offshore
was codenamed — he had predicted that the recent changed
procedures at the DDT 'could significantly increase our liability
exposure'.[8]

As Hardie focused on the US there was a growing realisation
by its managers that they should get a better handle on just how
much the 'asbestos legacy', as they called it, might end up
costing. The company's in-house lawyer had consulted the firms
handling its asbestos litigation in 1992 and reported his 'best
intuitive guess' of $40–45 million, but as the lawyers had agreed
'... it was like plucking figures from the clouds'.[9]

By 1996, Hardie commissioned the actuarial firm Trowbridge
to conduct a more accurate study of the likely claims. Trowbridge's
initial estimate of Hardie's exposure was $175 million, but within
four months its final estimate had risen to $230 million. More
alarming, the actuaries found that over three-quarters of all claims
against Hardie were for mesothelioma, a higher proportion than
other asbestos defendants, and the number was rising fast.[10]

Hardie executives had no intention of letting either the
market or its shareholders know about the secret looming debt.

In the next Trowbridge study in September 1998 the actuaries had increased their estimate to $254 million. Hardie's general counsel Peter Shafron instructed the firm to forward their report to Hardie's lawyers at Allens, marked 'Confidential and Privileged — for the purposes of Litigation'.[11]

In the broader investment community, the best that could be said was that nobody really had any idea how big the asbestos liabilities were. Hardie appeared unconcerned. A survey conducted by the company of opinions published by analysts reported simply: '... analyst estimates of the present value of the liability vary widely but appear to range from $50 million to $120 million.'[12]

However, the major public reason for the board's interest in relocating Hardie's headquarters was the taxation burden stemming from the growing structural imbalance between the company's shareholder base — firmly in Australia — and the site of most of its economic activity, the US. Hardie had minimised the tax paid on its dividends to Australian shareholders by routing them through a Hungarian holding company, but if it listed in the US that might no longer be possible.

The solution, the board was advised, might be to create a new holding company in the Netherlands, which could take advantage of a favourable US–Netherlands tax treaty, while Hardie's asbestos subsidiaries would remain as a rump under the Australian parent James Hardie Industries Limited (JHIL). JHIL would then steadily sell down its interest in the new Dutch company that would take over its role as the centre of the corporate empire.

Hardie's managers were instructed to develop the plan under the direction of a committee including the CEO Peter

Macdonald, chairman Alan McGregor and director Meredith Hellicar. Even at this stage, it was clear what their focus would be. 'Asbestos is the critical issue ...' began the report of the subsequent management meeting. Those present drew up a plan to better quantify the company's asbestos liabilities, mindful of the board committee's warning: 'There needs to be a strategy for managing Trowbridge and they need to be informed that parts of the report may be made public.'[13]

Secrecy continued to dominate the management preparations.

A study of strategies to shed their asbestos liabilities adopted by overseas asbestos companies, including Hardie's former partners Johns-Manville and Turner & Newall, gave little reassurance. US lawyers did offer the encouraging news about Hardie's former Canadian asbestos miner, Cassiar, that it had paid its creditors for a long time, then quietly gone into liquidation 'without a ripple'.[14]

As preparations for the move began in earnest, Hardie informed the US Securities & Exchange Commission that the structure of the new company to be floated on Wall Street was 'designed to insulate Newco from asbestos-related claims'.[15]

Baxter retained the advice of two high-powered PR companies, Gavin Anderson Kortlang and Hawker Britton, to assist him in developing a campaign which would win government and media support and neutralise opposition.

Asbestos was not Baxter's only worry. So were the tax-minimisation aspects of the plan: 'The proposed holding company structure and its relationship to an international financing subsidiary company will highlight the tax minimisation motives of the project. Tax is already firmly on the Australian political agenda as the country prepares for another federal election.'[16]

Hardie already had a reputation for sharp tax practices. The company in earlier days had purchased its imported raw asbestos through a company registered in the Bahamas, a world tax haven. Indeed, when John Reid was chairman, he had championed the duty of directors to pay as little company tax as was legally possible.[17] By 1989, an Australian Tax Office report cited Hardie as one of the country's fifteen biggest users of tax havens, receiving 78 per cent of its profits through them the previous year.[18]

It was common knowledge among analysts that the company went to extraordinary lengths to minimise tax during the 1980s, using the Channel Islands and even a subsidiary on the Toronto stock exchange. Elizabeth Knight commented in the *Australian*: 'Its use of tax havens is legendary and its tax-effective practices are never far from any assessment of the quality of James Hardie's earnings.'[19]

Baxter could foresee another image problem for Hardie in the media coverage: 'Based on Chelsea's [Hardie's] tax paying record, commentators will have an opportunity to debase the project in the minds of the general public and retail shareholders. This could place further pressure on government and regulators to act in response to concerns about our tax minimisation strategies.'[20]

But it was in relation to 'Apple' — the unmentionable asbestos — that Baxter devoted much of his attention. He knew from Hardie's polling that its 'Name Behind the Names' and 'Life Be In It' campaigns during the 1980s had all but erased the public's association between James Hardie and asbestos. Now, he warned, all that hard work could be undone: 'The separation of operating businesses from contingent liabilities has the potential to *recreate* a strong, public association between

Chelsea's [Hardie's] business and Apple. This could undo many years of repositioning work in Australia and New Zealand, and if it flows across the Pacific, taint the US business.'

The Hardie PR manager predicted that at the time of the restructuring announcement most journalists would know little about the company:

> ... but will develop a sudden interest without the appropriate background knowledge to put it into perspective. We will have to do this for them. We will need to educate them quickly while identifying and overcoming any prejudices or outdated perceptions they may have. For example, many will only remember Chelsea [Hardie] as synonymous with the old fashioned Apple-based products.

Baxter anticipated that on the day the restructure was announced the story would attract widespread interest from wire services like Reuters and AAP, who would most likely assign to the story junior reporters who had 'virtually no knowledge or understanding of even the most rudimentary facts about the company'. He suggested the resultant unpredictability in coverage might be mitigated by targeting 'key opinion-leading journalists' and providing swift responses to inquiries. Media training for the chairman and the managing director would be essential.

Another potential distribution loomed. What about Reid? Would he support the plan? He still exerted a major influence as Hardie's second largest shareholder. But the chairman, Alan McGregor was, after all, part of the 'family'. McGregor could come to the rescue.

We have not ruled out the possibility that members of the
'family' will publicly oppose or criticise the project as 'un-
Australian' and not in the best interests of the vast majority
of Australian resident shareholders ... We understand the
Chairman is going to convene a 'family' meeting in
Melbourne on the evening of the day of the announcement to
explain the rationale and merits of the project. We also
understand the Chairman is going to approach the former
Chairman a month before the announcement, with the aim of
bringing him into the loop and securing his early support.[21]

On 30 June 1998 the board gave the go-ahead. Hardie was all
set to move offshore.

A subsidiary company in Holland was renamed James Hardie
NV and took control of James Hardie & Coy's remaining
businesses, leaving the Australian manufacturer as a shell whose
only remaining activities were to lease its land and pay asbestos
claims. The parent company controlling the group remained the
Australian-based James Hardie Industries Limited (JHIL). The
next step was to create a separate vehicle for the US operations.
Papers for the float of the new company were lodged in New
York; by February 1999, Baxter was on a plane to the US, where
he met up with Peter Macdonald for a month-long road show
selling the package to investors. The Australian JHIL would
initially hold 85 per cent of the new US company; the remaining
15 per cent was to be offered on the New York Stock Exchange,
where Hardie's CEO would ring the bell at the opening of trade
on 5 March 1999.

But it was not to be. Timing was critical, and by the end of
the road show it was obvious the float was in trouble. Although

the stock was received favourably by many analysts, most US brokers believed it was overvalued. There was also concern about Hardie's stake in US gypsum operations, which were notoriously cyclical, as well as asbestos liabilities. Then on the day the public offering was announced, Russia defaulted on its sovereign debt, sending markets slumping, and in the week before the float the yield on thirty-year US treasury bonds began rising. Hardie's proposed stock was marked down by a jittery market.

The night before the planned float, the Hardie executives met dejectedly in New York and, after a long telephone hook-up, the board decided to scrap the launch.

Hardie was caught in a trap of its own making. With or without its New York float, the company's centre of gravity by 1999 was clearly in the US, where most of its revenue and profit was generated. And in Australia, a new accounting standard known as ED88 would soon force Hardie to publicly reveal a complete list of estimated current *and* future asbestos liabilities.

The company had only ever told shareholders that provision had been made for 'best estimate of known claims, which referred to the current year's asbestos claims that succeeded in court.

But what about the *unknown* claims? What was the total amount likely to be paid to people who may not even have fallen sick yet? Hardie simply asserted the figure was impossible to estimate: 'A contingent liability exists in respect of the ultimate cost of settlement of any claims to be made which cannot be measured reliably at this point in time ...'[22]

Could shareholders be given such a brushoff now that Hardie was receiving regular estimates from Trowbridge on precisely this 'ultimate cost'? Shafron thought so, describing the Trowbridge work as '... very uncertain ... it is based on very imperfect epidemiological models'[23] which could vary widely. History had shown, though, that such uncertainty with Trowbridge was always likely to be an *under*estimate.

But when the ED88 standard came into force, the estimates would have to be made public. What made it worse was that Hardie's competitor and fellow defendant, CSR, was planning to substantially increase provision for its liabilities in advance of the standard, leaving Hardie's asbestos secret looking increasingly threadbare.[24] Once the secret was out, Hardie's market value would probably plummet and a US float would become less likely than ever.

In Sydney, Allens' senior partner, Peter Cameron, who would play a key role in the restructure over the next few years, had outlined Hardie's asbestos choices in an opinion ominously entitled 'Ultimate Resolution'.[25]

Hardie could bring the issue to a head, he advised, by paying existing claimants and then winding up the asbestos companies, leaving future victims without compensation. Or it could provide a 'cushion' by establishing a compensation fund for future claimants which could be topped up with some extra cash as a 'safety margin'.

Following the abandonment of the US float, Hardie's general counsel Peter Shafron revisited Cameron's idea that the Hardie board could shed its asbestos subsidiaries altogether by placing them under the control of a separate foundation to administer compensation payments to claimants. Shafron, who had by now joined Macdonald in California, wrote a paper for the board's consideration, entitled 'Big Picture Options for James Hardie's

Asbestos Liabilities in Australia'.[26] Although Hardie had already split its company structure with the establishment of the Netherlands holding company, he observed, '... no recommendation was made nor accepted in respect to the end game'. The time had arrived for such an 'end game'.

Meredith Hellicar explained the board's reasoning: 'I think we hadn't realised how sensitive the United States was ... by that stage, of course, a whole raft of US companies had folded under their asbestos liabilities. So that certainly then started our thinking. Is there anything we could do? Would it be possible to separate out those liabilities?'[27]

She was quick to add that the directors never contemplated avoiding the liabilities altogether: 'Our sole aim was: how can we separate those liabilities from our balance sheet, properly provide for them, and would that make a difference to the rating of the company, in particular the US shareholders' perception of the company?'

According to Shafron there was no point waiting for the NSW government to change the Dust Diseases Tribunal. He argued: 'There is a strong institutional bias against James Hardie and other asbestos producers ... a single mistake by a defendant or a new document or finding serves as a kind of perennial fuel in its ritual immolation.'[28]

But the federal government was planning changes to Australian tax law that enabled an alternative strategy. Hardie's company in the Netherlands now ran all of the group's ongoing business, but was still owned by the parent, James Hardie Industries Limited in Australia, where most shareholders were based. The planned tax change would allow both companies to effect a share swap without incurring capital gains tax. Hardie shareholders could become the direct owners of the Netherlands company, leaving an empty JHIL in Australia responsible only for its asbestos subsidiaries.

Once again, Hardie executives focused on the idea of shedding the asbestos subsidiaries completely. They had little doubt that their most difficult task would be to neutralise objections from Hardie's asbestos victims and their lawyers. With the 'asbestos poison pill clearly separated from the operating assets', read a paper presented to the December board meeting, one of the major issues would be '... minimising the opportunity for asbestos-related spoilers to interfere and object'.[29]

By February 2000 Hardie's planning for what was now codenamed 'Project Green' was firming up, driven by Macdonald, who had been promoted to the group's CEO. Shafron took some additional soundings about 'hiving off' the asbestos companies. He approached Michael Gill, a senior partner at the Sydney law firm Phillips Fox, whose involvement was to remain secret.

In a telephone hook-up from California, Shafron made it clear to Gill that at this stage the company intended to provide for everyone who became sick from asbestos, including those in the future. But he was adamant that once the foundation was established, it had to survive on the money it was given. There would be no top-ups.[30] This was, indeed, the end game.

Gill was alert to public sensibilities. The media would be interested, he warned, and stakeholders would be suspicious. They would need to be convinced that it was a good deal. 'How to explain the uplift ... on *60 Minutes*?' he mused. He advised Shafron that the best way for Hardie to achieve an 'unbridgeable corporate split' with its asbestos subsidiaries would be to follow a course involving the least public scrutiny: 'It will be perceived as a big corporation attempting to squeeze out of its responsibilities to the 'victims' of its corporate activities.'[31]

Shafron had commissioned another estimate from Trowbridge and was shocked when the preliminary report arrived on his desk in California in March 2000. Hardie's asbestos liabilities had risen 40 per cent to $300–350 million, compared to the previous year's figure of $254 million.

'Wow. That's much more than I was expecting,' he emailed Trowbridge.[32]

The reason, explained Trowbridge's actuary David Minty, was that a new ruling allowed asbestos claimants to be reimbursed for the cost of work around the house they could no longer perform themselves. He sneered, 'They were all brilliant mechanics, bricklayers, carpenters, painters and gardeners ...'[33] As a result, the average cost of a mesothelioma settlement had increased by 10–15 per cent. Newly eligible claims from waterside workers had also been factored in at a cost of $8 million.

Hardie's asbestos litigation lawyer, Wayne Attrill, showed for the first time some realisation that the company's recent hardline legal strategy might have been counterproductive. He emailed Shafron: 'Yes, Peter, this is definitely a problem. As you know, I had thought that JH's investment with Allens in working up the test cases would yield us long-term dividends in the form of lower settlements. This has proven not to be the case.'[34]

Planning on the proposed restructure intensified, with Shafron instructing those involved to maintain '... the strictest levels of security', urging them to mark their documents 'Private & Confidential' and to shred or delete those which were not essential.[35]

He focused his attention on the latest Trowbridge estimate, urging the Sydney-based Attrill to 'stay close' to the actuaries, and insisting that the report in progress must stay just that — a work in progress, without the finality that would impose legal significance. 'Don't let them go final whatever you do,' he instructed.[36]

Trowbridge slightly moderated its earlier prediction down to $293 million, but Shafron was not content. He demanded the deletion of references to earlier reports and asked the actuaries to 'tone down speculative risks'.[37] Other key changes were made at his request. The sentence 'Wide variations are normal and to be expected' was deleted, along with the word 'considerably' in the warning 'future experience could vary considerably from our estimates'. Trowbridge drew the line when Shafron asked for a 'cosmetic' change by deleting its sensitivity analysis, the section drawing attention to the variability of the estimate if any key assumptions were to change. The dispute was never resolved. The report remained a draft, never signed off as a completed document.

But Shafron was sufficiently appeased by the alterations to use the report to shop around for insurance against the asbestos liability, despite Trowbridge's misgivings. The search proved fruitless, with potential insurers far more pessimistic than Trowbridge. AIG gave the actuarial study short shrift, increasing its estimate of Hardie's undiscounted liability by more than $100 million. The cost quoted for asbestos insurance was prohibitive: a premium of around $400 million for a total liability of $1 billion.[38]

What if the Trowbridge estimate was wrong and the foundation would in fact need more money? At least one of Hardie's lawyers at Allens thought Trowbridge had been 'excessively optimistic'.[39] The estimate had not taken into account, for example, the likelihood of the courts setting new precedents and awarding higher damages, nor had it included an increase in payouts for lung cancer. Shafron dismissed this view as 'gloom and doom'.[40] He was reassured by Allens' senior partner, Peter Cameron, who advised that, although 'politically' an additional buffer might be worth considering, Hardie directors would discharge their legal duty as long as they provided the foundation with the amount the actuaries estimated would be needed.[41]

The Trowbridge report remained heavily qualified even after Shafron's alterations. In its calculations Hardie most commonly quoted Trowbridge's median range, which the actuaries warned had only a fifty–fifty likelihood of accuracy. In fact, the estimates were still so littered with disclaimers that Hardie's spinners doubted their usefulness. Steve Ashe, who had assisted Baxter in developing a communications strategy to neutralise possible opposition, expressed his frustration: 'It does not leave the reader with confidence that the amount of $294m is sufficient. In fact, one could easily be left with the impression that the amount is insufficient.'[42]

Ashe warned that there were a number of potential lines of attack against Hardie's plan:

- Now that the company appears to be doing well it wants to preclude access to victims (and their families) upon whom they have inflicted the asbestos-related disease ...
- Past experience suggests that JH is a company that cannot be trusted. The company has a history of not disclosing the truth — for its own gain ... they have done it before — and are doing it again.

- Their behaviour in court provides ample evidence that they will do whatever they can to avoid or minimise the amount paid to victims ...[43]

The following month Hardie approached an alternative actuary who provided even less satisfaction. The firm was reluctant to choose a best estimate, tartly warning Hardie that anything other than a range of liabilities would be of 'spurious accuracy'.[44] Hardie quickly abandoned the attempt and stuck with Trowbridge, where the news was to get worse.

At an industry conference in November 2000 two Trowbridge actuaries cautioned that the peak of mesotheliomas in Australia had not yet occurred and that '... many insurers and other parties exposed to asbestos-related disease liabilities may be significantly under-reserved'. They had also factored in the possibility of increased claims for lung cancer and a trend towards higher judicial awards.[45]

None of these estimates came as a surprise to Shafron, even though it meant an increase of around 40 per cent to Trowbridge's March estimate of the company's liabilities. He emailed Hardie's CEO Peter Macdonald, confirming that the prediction '... broadly accords with our own experience'.[46]

But what did greatly concern him was the prospect of this information becoming publicly known. He immediately asked Hardie's Sydney lawyer Wayne Attrill if Trowbridge would remove the new figures from its website. As he warned: '... their broad public release could well attract wider attention ... we were very surprised to hear of the report, given that we have Trowbridge on a retainer on this very subject ... Peter Mac hit the roof when he saw the report.'[47]

The significance of Shafron's and Macdonald's concern was not that the estimates were getting worse; it was that others might hear about it.

Trowbridge refused to remove its web page, but Attrill consoled Shafron with the observation that although the conference where the figures had been released was well attended, it was unlikely to be reported. 'No media,' he emailed.[48]

Media, in fact, were only one of many potential problems facing Baxter and his PR team in their preparations to sell Hardie's restructure. As he workshopped the 'stakeholder' issues, he had no illusions that it would be a difficult task, warning that the board should be prepared for a fight.

Top of the list of threats was possible intervention by the NSW government, under pressure from unions, asbestos victims groups and plaintiff lawyers. Although Hardie's 1998 lobbying effort to block changes to the Dust Diseases Tribunal had ended in an embarrassing disaster, it had recently rehired the PR agency Hawker Britton to renew pressure for change to the compensation system. Work to build bridges with the government appeared to have borne fruit, and an adviser to John Della Bosca, the minister who dealt with asbestos diseases, had offered to provide early warning on issues of concern.

Just as they had done two years earlier, Baxter and his team began drawing up a list of people who would need to be lobbied and the talking points they could use. Would the proposed foundation have sufficient funds?

Baxter's draft background notes for the media launch never contemplated the prospect that it could run out of money: 'In the event that the investment income is insufficient to fund liabilities as they arise, Trustco will draw on its significant capital base to meet any such obligation, until there are no further claims.'[49]

This reassuring line was to subtly change as the shape of the final arrangements emerged. But even at this early stage Baxter's notes would have alarmed asbestos victims groups, if they had known their contents. He predicted that '... claims will continue for 10–15 years', yet even the Trowbridge projections made it clear that people would continue to fall sick from the company's use of asbestos well beyond then.

Shafron was also worried about how to defuse the arguments of potential 'spoilers' to the restructure. A plan crystallised in his mind that might disarm their objections. If James Hardie Industries Limited, the original parent company in Australia, were to retain a capital lifeline through to its Netherlands replacement, JHI NV, then the status quo could effectively be preserved. In exchange for the transfer of assets to the overseas company, JHIL could issue its Netherlands counterpart with partly paid shares equal to the asset worth of $1.5 billion. Then anytime JHIL wished, it could demand that the Netherlands company pay the full value of the shares. In this way the two companies would swap shares, but JHIL would be left with access to the same amount of capital it had at the start.

'Stakeholder concerns virtually disappear,' Shafron wrote excitedly to his colleagues.[50] If the stakeholders were appeased there would be less cause to worry, nor would Hardie need the actuarial opinion. As a bonus, there was nothing to stop the board deciding later to sell JHIL's shares, declare a trust over them, or distribute them to the ultimate shareholders if it wanted to.

A Hardie finance officer objected. What if asbestos victims could, after all, sue the Australian parent company? How secure really was the corporate veil that so far had protected it from suits incurred by the former giant asbestos manufacturing subsidiary, James Hardie & Coy?

> My concern is that, with this partly paid investment in place, should circumstances arise where JHIL was found to be liable for JH & Coy's liabilities (I grant you that this is certainly not the case at present, but who knows what legislative or other changes may occur in the future), the directors of JHIL would have little alternative but to make calls upon the unpaid capital.[51]

The director from Hardie's biggest shareholder, Brierley Investments, echoed his concern. But the financial exposure which worried both men would only remain a problem while the lifeline of partly paid shares remained in place. A fortnight later Shafron sought advice from Allens about the possibility of cancelling the shares: 'What if the company wished to give thought, in the light of circumstances obtaining in the future, to cancelling some or all of the partly paid shares? What then?'[52]

As 2000 drew to a close, preparations to put the board strategy into effect gathered pace. A company to assume the role of foundation and take control of the asbestos subsidiaries had to be created and directors found. The foundation would need enough money to convince its new directors as well as the government of its viability. Hardie had finally struck a deal with QBE, which had agreed to pay Hardie $47 million for the next fourteen years. But the two former asbestos subsidiaries, James Hardie & Coy and Jsekarb, the former brake company, could still only scrape together less than $200 million, nowhere near enough to match the problematic Trowbridge estimate. They would need more.

Costings were made on the possible earnings of the foundation, based on the income it might receive from the sale of properties, rental of factory sites back to Hardie and other investments. The same Hardie finance officer estimated an annual earnings rate for the foundation's cash flow at a highly optimistic 11.7 per cent, a figure soon to be queried even by Hardie consultants. The finance officer later joked about the 'well-loved financial model' with a rate '… we used to convince the board, the foundation, its insurers and indeed ourselves of the financial outcome'.[53]

As the year drew to a close, Shafron and Macdonald set about the delicate task of recruiting trust directors. They had already decided on a suitable chairman. After ten years on the Hardie board, Sir Llew Edwards was tiring of the long trips to California. Hardie's chairman Alan McGregor had already sounded him out to head the asbestos foundation and he seemed keen. His medical qualifications were an additional bonus for a foundation whose minor role would be to sponsor research into asbestos diseases and possible cures. At $90,000 a year for six board meetings, it was an attractive prospect. Sir Llew accepted the invitation and would soon retire from the Hardie board.

A suitable candidate for managing director, it transpired, was already under their noses in California. Dennis Cooper, the company's IT specialist, had been relocated from Australia during the first aborted plan to float on the New York exchange, and was looking forward to returning home. Shafron approached him in the office, explaining the company's intention and leaving Cooper a week to mull over the idea. He would earn the same as Edwards, and he figured he could do the job in three days a week, perfect for the part-time role he'd planned when he got back to Sydney. He, too, said yes.[54]

'Phew!' emailed Shafron. 'Got one on the hook,' he later told others in the management team.[55]

Next was the Sydney-based Phillips Fox lawyer, Michael Gill, who had provided Hardie with advice about achieving 'an unbridgeable corporate split'. He received a message that Hardie required a 'worthy' for the board of the new structure, and he succumbed to Shafron's personal approach.

For the final choice, Shafron and Macdonald selected the only candidate who was truly independent from James Hardie, Peter Jollie, a former president of the Institute of Chartered Accountants. Extraordinary care was taken to entice him into swallowing the bait. Macdonald faxed through a finely crafted script for Edwards to read to him on the phone.

'Good day. I would like to sound you out on a potential role as a non-executive director. You have been suggested by our advisers as a person of suitable background and experience to be considered for an important director role of a trust to be established by James Hardie ...'[56]

Jollie accepted, a decision he would live to regret.

Back at Hardie's operational headquarters in California, Macdonald and Shafron were almost ready. All they required was board approval of the deal and Hardie could 'leave guilt behind', as one of the management team put it.[57] The next challenge would be to launch the trust with the minimum of publicity. Once the foundation had been launched and accepted, Hardie could embark on the second stage of its 'end game' and move its parent company offshore.

ten

THE 'FULLY FUNDED' FOUNDATION

When Hardie's PR chief Greg Baxter first explained the foundation Hardie had decided would house its former asbestos subsidiaries to his colleague 'LJ' Loch, then with Hawker Britton, it wasn't long before Loch asked the critical questions, starting with: what happens when the money runs out?

Baxter's answer was blunt:

BAXTER: It runs out.

LOCH: What guarantees can JH [James Hardie] give that victims aren't going to be left stranded?

BAXTER: None.

LOCH: They are the two big questions?

BAXTER: Agree …

LOCH: What's in this for JH? How do we know it isn't about JH pulling a 'swifty' and walking away from its responsibilities here?

BAXTER: Let's discuss our position on this.[1]

Baxter had no ready reply. They obviously had more work to do. Loch considered the implications:

> LOCH: What happens 20 years from now when the trust
> runs out of money if there are still people who are dying as a
> result of JH products? Can JH be made to top up the fund?
> BAXTER: Our legal advice — silk department — is that the
> liability is confined to the subsidiaries …

Hardie's legal advice had, in fact, been mixed. While a recent attempt by New Zealand worker Desmond Putt to break through the corporate veil and sue the parent, James Hardie Industries Limited, had failed in the High Court, new precedents could always be established. The High Court had even hinted as much in Putt's case.[2] The company's own asbestos litigation manager wrote there was a 'real risk' that this line of attack would be used again in the future and might ultimately prove successful. He added that this was a risk being addressed by the plan to hive off the asbestos subsidiaries to a new foundation and then shift ownership of Hardie's operations to a new parent company in the Netherlands.[3]

But this risk was not one that Baxter felt obliged to explain to Loch, who remained troubled by the idea that the foundation could run out of money:

> LOCH: What can JH do to cover off on the 'what if'
> question? There must be some sort of certainty it can give?
> BAXTER: There isn't.

For Loch and her Hawker Britton colleagues, Baxter's answer posed a major problem. She emailed Baxter a summary of their discussion, with the rider:

LOCH: Greg, the big question which we all kept coming back to was the guarantee question. Government will need some level of comfort that this isn't going to leave 'victims stranded'.
BAXTER: There is no guarantee. And can't be, even if their subsidiaries stayed inside JH.

This candid exchange between Baxter and Loch set alarm bells ringing in the head of David Britton, a principal of the PR firm.[4] But for Baxter, it was simply a statement of how the law then stood. Provided Hardie was protected by the corporate veil, if its asbestos subsidiaries ran out of money there would be no more compensation payments.

Baxter had no illusions about the difficulty of selling his message. He had pithily summed up the challenge to the Hardie board: 'Our central communications conundrum is that we will not be able to provide key external stakeholders with any certainty that the funds set aside to compensate victims of asbestos diseases will be sufficient to meet all future claims.'[5] Yet within six weeks 'certainty' was exactly what Hardie would promise.

Baxter's colleague, Hardie's general counsel Peter Shafron, already knew that the two JHIL Australian asbestos subsidiaries were worth less than the latest estimate for future claims. This situation had been made worse because the larger subsidiary, the asbestos cement manufacturer James Hardie & Coy, had been gutted of assets.

Shafron thought one solution to this lack of funds might be to reverse a dividend, then worth more than $57 million, which Coy had paid its parent JHIL in 1996. Even when it was paid, future litigants might argue that the parent board should have known that it might leave Coy without enough money for its asbestos liabilities, because the dividend was issued a few days after the company had received its first draft estimate from Trowbridge of future claims.

The idea of repaying this dividend appealed to Hardie's chairman, Alan McGregor, who felt it might make the foundation 'more palatable' to the directors Hardie had chosen to run it. All of these directors, with the exception of Peter Jollie, had a strong connection to Hardie, despite the intention to promote the new foundation as independent. Sir Llew Edwards had been on the board of James Hardie Industries for over a decade; Cooper had been a senior employee for more than five years; and Gill, who had advised Hardie on the restructure, was insisting that the foundation must have enough funds to last fifteen years.

Shafron also had another proposal, which he emailed to CEO Peter Macdonald. If JHIL topped up the foundation's money by refunding the dividend, then in return it could slip an indemnity clause into the arrangement against any future asbestos suits: 'Obtaining the indemnity overcomes possibly the biggest question mark I have over this transaction (risk to JHIL) — I would very much like to make it work.'[6]

But to make the deal work, the new directors of the foundation would have to be given details of the expected asbestos liabilities. Another Trowbridge report was needed. Hardie's lawyers at Allens suggested that it should be commissioned by the foundation's lawyers (whom Hardie had chosen and was paying for), but Shafron disagreed, emailing back: 'I want the report to be JHIL. I want to keep Minty [the Trowbridge actuary] on the JHIL side of things as far as possible, for tactical reasons and control.'[7]

He instructed Trowbridge to speedily prepare another, briefer report, based on its March 2000 study and projecting ahead no more than twenty years. The shut-off date for input of new asbestos claims data into the new model was critical. Hardie had received claims for the next nine months which would not be included in the model. Trowbridge's David Minty later claimed that Hardie blocked access to the latest claims data with the excuse that its computer specialist was on leave; Shafron insisted that the actuaries told him they did not need the latest figures because '... it would be unlikely than an additional short period of data would make much difference'.[8]

In fact, the missing months made a big difference because they reflected a dramatically worsening trend, with the cost of mesothelioma settlements almost doubling. Without the new claims data the Trowbridge estimate was likely to be 'grossly inaccurate', Hardie's in-house asbestos lawyer Wayne Attrill later admitted.[9] He had already put in a bid to run the foundation's litigation as an outsourced consultancy.

When Trowbridge's rushed and abbreviated report was completed its median best estimate for Hardie's twenty-year liability was $287 million.[10] (This figure proved to be a massive underestimate, which within six months Trowbridge would revise upwards by a huge 65 per cent.) However, even with this estimate, Baxter believed that the Hardie directors needed 'comfort on numbers' to soften their 'moral reservations' to the restructure.[11] In return for an indemnity against asbestos suits, Hardie would pay an additional $80 million into the foundation's assets in yearly instalments of $12.5 million, with no reference to the dividend payment that had inspired the extra money. Combined with existing assets and Hardie's 'wildly optimistic'[12] cash-flow model, the foundation should have just enough money to meet the new Trowbridge estimate.

But would the new figure be enough to convince the foundation's new directors? Hardie's chairman reported back to his board that the foundation's money supply was a potential 'dealbreaker'.[13] Cooper, the handpicked managing director, had already asked to see copies of previous Trowbridge reports. He was told Shafron would have to clear his request and never heard anything more. Gill, as the only insurance lawyer on the foundation board, felt his colleagues relied on his advice and sought talks directly with Trowbridge before the first meeting of the new board. Shafron, however, stalled these talks until he could be present. He informed his colleagues that all the prospective directors seemed to be 'in the slot'.[14]

When the Trowbridge actuary David Minty eventually did brief the foundation, he realised afterwards that the directors may not have known that his projections were based on the condensed version of his old study. He discussed the matter with a colleague on the way back from the meeting and agreed to add the words 'March 2000' for the next presentation.[15] At the subsequent foundation board meeting the change went unnoticed by the new directors, but two days later a young Allens lawyer did spot it and worried about the implications.

It was 15 February 2001, the day chosen for the board meetings of both Hardie and the foundation to approve the scheme. Like the new foundation directors, Hardie adviser David Robb from Allens had assumed that the 'updated' Trowbridge report included the most recent Hardie's claims data. That morning Robb noticed the addition of 'March 2000' to the Trowbridge title page. Alarmed, he called Shafron, who reassured him that Trowbridge believed the new data was not necessary. Robb was still not satisfied and voiced his concern to his senior partner, Peter Cameron. They phoned Hardie's CEO Peter Macdonald in California, who repeated the assurance that Trowbridge had said it did not need the latest figures. Cameron cut to the chase: was there any reason to depart from the view that the foundation would be fully funded? 'Absolutely not,' said Macdonald.[16]

The remainder of the day went according to the 'aggressive' script prepared by Allens and Shafron. Hardie directors approved what seemed 'a perfect marriage of good business sense and good corporate social responsibility', as Meredith Hellicar later described it. She said she had no doubt about the funding:

> We were a much poorer company in those days … had we
> known the correct figure in those days, we wouldn't have
> done it. We wouldn't have done it, but, of course, more
> importantly we might not have survived. I mean, if we'd had
> to bring that figure onto the books, presumably, I don't
> know what we would have done, but presumably we would
> have been in danger of folding ourselves.[17]

At Allens' headquarters in Chifley Plaza the foundation directors were asked to sign a steady stream of papers. For the new managing director, Dennis Cooper, there was a blur of people coming and going from different rooms.

> My overriding memory was one of being programmed by
> Allens, who were essentially dictating all of the events that

would occur ... it was like a very complex film or drama,
where you stay engaged for the first three subplots and then
suddenly your mind goes a little bit blank as you move into
the fourth, fifth, sixth, seventh, eighth ... [18]

The foundation board meeting became an all-night session.
During the seven-hour process, Cooper was confident he could
trust those who were driving it. For him it was unthinkable that
there would not be enough money for the foundation in the
future. Hardie's CEO Peter Macdonald reassured the directors
several times that night: 'There will be no better friend to the
foundation than James Hardie,' Cooper recalled him promising.
'We're going to make it work.'[19]

Baxter planned to announce the news the following day. Like
Hardie spinners before him, he had decided to bury it as a
business story with the aim of 'attracting as little attention as
possible'. The announcement would be made to a combined
briefing of financial analysts and business media at the Friday
presentation of Hardie's third-quarter results.

'Our most desired outcome,' he noted, 'is to have the analysts
walk away from the presentation understanding that the
establishment of the trust means that JH no longer has any
significant liability for asbestos.'

A tight morning schedule and the abandonment of the usual
webcast ('we would explain this as due to technical difficulties')
would minimise the risk of other media 'hijacking' the briefing.[20]

Again like Hardie spinners of old, Baxter had rehearsed the
media lines of the two board chairmen and CEO during mock
interview sessions in the wood-panelled boardroom of the old
Asbestos House. A TV crew and journalist were hired to add

realism as Macdonald, McGregor and Edwards practised their
finely honed scripts:

Q: What if the foundation runs out of money?
A: James Hardie is satisfied that the foundation will have
sufficient funds to meet anticipated future claims ...

Much of Baxter's preparation focused on ensuring a smooth
reception of the scheme by the NSW government. Neither the
federal nor any of the state parliaments would be sitting that week.
He rated the chance of legislative intervention as low. NSW Premier
Bob Carr, he predicted, would most probably 'flick-pass' the issue
to Canberra, where the Liberal prime minister, John Howard, held
constitutional power over corporations. Another 'relatively
meaningless' option, in Baxter's view, for Carr's government to
'take the political heat out of the issue' would be to establish a
government inquiry '... to 'show' it was doing something'.[21]

Carr, his chief of staff, Graeme Wedderburn, and Cabinet
Secretary Roger Wilkins were identified as key figures in
Hardie's lobbying strategy: the premier was 'pro-business' but
sensitive to 'political exposure or a risk to his government'; his
chief of staff did not want the issue '... to become a 'cause' for
tabloid media'; and Wilkins 'would not welcome debate about
corporate veil issues'. The industrial relations minister, John
Della Bosca, would probably focus on whether the foundation
had enough money. Bob Debus, the attorney-general, was all but
written off in Baxter's strategy document: 'Aligned to the more
militant factions ... the most likely ... to suspect JH's motives.'[22]

There had been an unexpected hitch in Hardie's
communications strategy. After Baxter's workshopping session

with Hawker Britton, one of the company's principals, David Britton, had taken fright. He told Hardie the firm would no longer do the job, after being warned off by his close friend Greg Combet, the secretary of the Australian Council of Trade Unions, who had not minced words: 'Keep an eye on those James Hardie bastards! [If they] try to wriggle out of their asbestos liabilities [I'll] come down on them like a ton of bricks!'[23]

Baxter quickly sought advice from Hardie's other PR consultant, Gavin Anderson and Company,[24] which suggested hiring Stephen Loosley, a former senator and secretary of the NSW Labor Party, and then a consultant with PricewaterhouseCoopers (PwC), where it was expected he would soon be joined by the former national secretary of the Labor Party, Gary Gray.

Loosley accepted the Hardie brief, which during the next three months would earn him nearly $50,000. He even resumed the consultancy three years later at a time when most political figures were scrambling to put as much distance between themselves and Hardie as possible.[25]

To add credibility to Hardie's plan, Loosley suggested the company seek independent endorsements of the foundation's funding. Hardie asked PricewaterhouseCoopers and Access Economics to 'bless' the foundation's cash-flow model, an exercise a company executive later agreed was actually 'arid and pointless'.[26] When both firms queried the model's optimistic earnings rate of 11.7 per cent, Hardie told them it wanted them to check the technical competence of the model, not comment on its assumptions. Access still cautioned that the earnings rate warranted '... detailed consideration by James Hardie'; PwC echoed the warning: 'We urge the directors of James Hardie to satisfy themselves as to whether the values and assumptions used in the model are reasonable.'[27]

Loosley helped to organise a briefing with Carr and Della Bosca's chiefs of staff, and also suggested broadening the profile of the foundation board, offering to make a private approach to Hazel Hawke, the popular ex-wife of Bob Hawke, and Lady Martin, the widow of the former NSW governor Sir David Martin, who had died from mesothelioma.[28] The briefing with the two public servants went better than expected. Hardie's notes from the meeting even record there was a hint by Graeme Wedderburn that the government would like to change the asbestos compensation regime in order 'to curb ambulance-chasing lawyers'.[29]

When Baxter's media plan sprang into action, no press conference was held: 'We will deal with general media one-on-one. This will help us avoid a media siege ...' His overall aim was to position the foundation as 'a credible, independent organisation ... able to take the moral high ground in ways denied to JH ... focused purely on compensating asbestos disease sufferers with genuine claims ...'[30]

Hardie's media release, lodged with the stock exchange, was unequivocal about the prospects of future claimants. Headlined JAMES HARDIE RESOLVES ITS ASBESTOS LIABILITY FAVOURABLY FOR CLAIMANTS AND SHAREHOLDERS, it proudly announced the formation of the 'completely independent' Medical Research and Compensation Foundation and declared:

> The foundation has sufficient funds to meet all legitimate compensation claims anticipated from people injured by asbestos products that were manufactured in the past by two former subsidiaries of JHIL.

JHIL CEO, Mr Peter Macdonald said that the
establishment of a fully funded foundation provided
certainty for both claimants and shareholders.[31]

In case anyone missed it, the word 'certainty' was repeated in the
next paragraph. A later government inquiry found the release to
be 'a pure public relations construct, bereft of substantial truth',
because the foundation was, in fact, 'massively under-funded'.[32]
But Hardie's release stated the opposite, invoking the names of its
confined consultants to imply their endorsement:

In establishing the foundation, James Hardie sought
expert advice from a number of firms, including
PricewaterhouseCoopers, Access Economics and the
actuarial firm Trowbridge. With this advice, supplementing
the company's long experience in the area of asbestos, the
directors of JHIL determined the level of funding required by
the foundation.

As hoped, business media showered praise on the restructure,
complimenting Hardie for at last '... shedding the baggage of
the past'.[33]

One critical voice was journalist Ben Hills, who had followed
the asbestos story since he tracked CSR's behaviour at
Wittenoom in the 1980s. In the *Sydney Morning Herald* he
quoted the plaintiff lawyer Armando Gardiman: 'My concern
would be — what happens to the victims when the $293 million
runs out?'[34] This concern was echoed in a subsequent Hills article
by another plaintiff lawyer, Peter Gordon: 'If they intend to close
down those companies and then say they have no assets to

compensate the victims, I will be down on them like a wolf on the fold.'[35]

Greg Baxter drafted a response to Hills from Hardie's CEO Macdonald, asserting that 'The company and I have been very open and transparent in our disclosures about the foundation'.[36]

But behind the scenes Hardie's counsel Peter Shafron had emailed Trowbridge cautioning the actuaries not to speak to Hills.[37]

Macdonald's response repeated the claim that '... James Hardie (JHIL) believes that all valid claims for compensation will be met', a view based on the advice the company had received from 'expert and independent' advisers, including PricewaterhouseCoopers, Access Economics and Allens.

A lawyer at Allens, concerned about the firm's association with the false claim, emailed his objection to Shafron: '... a possible interpretation of Greg's letter is that Allens ... confirmed that the fund is sufficient to meet all valid claims. That is not so with respect to Allens and ... we would be grateful if special care were taken in future to avoid any such impression being given.'[38]

Shafron soon shut him up: 'I am a little concerned by the tone of your email. We kept Allens close by our side throughout this transaction (and Allens' billings will reflect that).'[39]

Wayne Attrill, Hardie's own asbestos litigation lawyer, claimed later that he, too, had expressed concern to Baxter soon after Hardie issued its press release.

'I just didn't think that that could possibly be said in such categorical terms and I expressed that view to corporate affairs. He said, 'Oh no, we're comfortable with that.'[40]

Baxter said neither he nor anyone else remembered Attrill voicing this criticism.[41]

* * *

During the days that followed the launch Baxter and Macdonald worked tirelessly to ensure the foundation was received favourably, briefing those they had identified as key stakeholders.

The reaction of NSW Industrial Relations Minister John Della Bosca could hardly have been more encouraging. He released a statement in which he 'cautiously welcomed' Hardie's announcement as '... an important acknowledgement of the company's responsibility to its former workers'.[42] Della Bosca pledged to work with unions, victims and Hardie to 'ensure the proposed financial and corporate structures are sound'. This was to prove a hollow promise and the statement was later removed from the minister's website.

Hardie also met with Michael Costa, the secretary of the NSW Trades and Labour Council, who voiced scepticism about the foundation's funding, but never took up Baxter's offer of access to Trowbridge's updated study. Macdonald told his Hardie colleagues that it was a matter of quickly hammering home the company's advantage to 'put the issue 'to bed' as soon as possible'.[43]

But Paul Bastian, NSW state secretary of the metal workers' union, wrote urgent letters to Carr, Della Bosca and Attorney-General Bob Debus, posing some 'unanswered questions' about the foundation. Had the government seen Hardie's actuarial assessment? 'In the worst case scenario,' he asked, 'if the fund exhausted itself in say 10 or 15 years, what does James Hardie say is its moral and legal responsibility to asbestos victims thereafter?' Would the parent company honour future liabilities?[44]

Bastian sent three copies to the premier — one posted, one faxed and the third hand-delivered to his office. Carr later denied receiving any of them: 'It's an absolute mystery! An absolute mystery. I said to the head of the Cabinet Office, Roger Wilkins, 'Our credibility rides on this, where is this document?' And there was a comprehensive search of the premier's office and the Cabinet.'[45]

The letter was never found in any of its forms and Carr remained adamant that he was never warned by the union about the Hardie scheme, despite a succession of criticisms which appeared in the media during the following weeks.[46] The union's complaints appeared to fall on deaf ears. Della Bosca spoke positively about the foundation in parliament, asserting that its capacity to pay would be 'vigorously tested'. Hardie was gratified, with Baxter's colleague Ashe emailing Loosley that the minister's comments were 'very good'.[47]

Hardie had nothing to fear from the 'vigorous testing' Della Bosca had in mind that was to take place at a meeting at the Dust Diseases Board. Hardie insisted the AMWU attend this meeting without its lawyer, arguing that '... legal representation for any party is neither required [n]or desirable'.[48] At the meeting, the union's representative demanded answers to a list of questions, but Hardie was dismissive. Afterwards, Baxter reported with satisfaction, the board's employer representative felt key concerns had been 'largely addressed', a view echoed by the executive director, who told Baxter he '... didn't see any reason why Della Bosca should be critical of JH or the foundation'.[49] This meeting was to be the limit of government scrutiny.

Nor was the Australian Council of Trade Unions (ACTU) in Melbourne to prove an impediment when Baxter and Macdonald met with its assistant secretary Bill Mansfield, who stood in for Greg Combet, the ACTU secretary. Combet later claimed that Mansfield 'asked all the right questions'.[50] In fact, Mansfield said he was left with the clear impression that if the money ran out, Hardie would make it up: 'If the funds ran out and there was legal action to secure additional money from James Hardie that was successful, I took them to be saying their liability would be met.'[51]

Yet Hardie's advice was that such legal action against the parent company was unlikely to succeed. To make absolutely sure, it had secured an indemnity from its former asbestos subsidiaries against such claims.

Baxter cynically summarised his meeting with Mansfield:

> He concluded the meeting with a summary of the key issues from his perspective as follows:
>
> - 'on the question of whether we can get access to the actuarial numbers, I am hearing that we can'
> - 'on the question of whether other JH assets are available for compensation if the foundation's funds are exhausted, I'm hearing that JH believes the liability is restricted to the two companies in the foundation'
>
> He's wrong on the first point and right on the second.[52]

Looking back, Combet conceded the ACTU should have done more: 'Perhaps we can all be justifiably criticised for not investigating it more. Was there enough money? We were terribly apprehensive that there wasn't. We'd received an assurance that there'd be more money if it was necessary. Could you trust them? Well, fucked if I know. In that context and at that time you wouldn't have thought so.'[53]

In Sydney, the AMWU continued to agitate. It organised a demonstration outside the former Asbestos House.

Hardie responded with a bold media statement unambiguously promising that the foundation would meet all future claims: 'This certainty for claimants, not possible under the previous structure, represents a significant breakthrough in the area of asbestos diseases compensation.'[54]

But as the demonstrators conducted their small protest outside the Hardie headquarters, the future of the 'independent' foundation housed inside the same building was already uncertain. Within six weeks of its creation, Dennis Cooper, the new managing director, realised that it might not have enough money. A quiet, trusting man, Cooper had been looking forward to an easy life in transition to retirement, but his fateful decision to take the job was 'to change my life', he said later. 'It was to be four years of hell.'[55]

As Cooper prepared the foundation's budget and first annual report, he compared his figures with the cash projections provided by Hardie. He was confused. His numbers were significantly higher than Hardie had forecast. The following month he asked Wayne Attrill, the former Hardie lawyer now contracted to run the foundation's asbestos litigation, to explain.

'It was only then that I realised that those numbers on the forecast spreadsheet were from March 2000 and had not been updated in any way by the experience of the nine months to December.'[56]

Despite these early concerns, Cooper told me he felt no 'panic'. He wasn't aware that in the same building the Hardie management team was already secretly planning the next move of its corporate restructure. Hardie's general counsel Peter

Shafron had urged his colleagues to 'thread the needle' to ensure that 'sensitive' documents were concealed from the foundation's view. Not to do so, he emailed, '... will likely make things tense with the new Board, who will become suspicious'.[57]

The foundation's directors had shown no signs of suspicion. Chairman Sir Llew Edwards was chancellor of the University of Queensland and had extensive experience on the boards of James Hardie, Westpac and other blue-chip enterprises. Cooper reasoned that if Sir Llew was relaxed, he should be too. But he experienced a growing anxiety. The new claims figures all told the same story: the foundation's finances were worsening at a far faster rate than Hardie had predicted.

Cooper buttonholed Peter Macdonald on one of his frequent fleeting trips though Sydney about this 'sensitive issue', but the Hardie CEO was breezily reassuring, saying the trend had been consistent for the past six years and that it would take several quarters — 'most likely years' — of significantly different numbers for the picture to change.[58]

Macdonald provided far less consolation when both Cooper and Sir Llew snared him for a meeting in mid-May and warned him that if the trend in foundation finances continued, '... we are going to be out of business in a very short time'.[59] Cooper attempted to give to the Hardie CEO a spreadsheet he had prepared revealing the inadequacy of Hardie's forecasts. Macdonald made no move to take them, keeping his hands at his side and dismissing the need to read them as neither necessary nor appropriate.[60]

Cooper said he felt momentarily embarrassed. The meeting spluttered to a quick conclusion, leaving him concerned and bemused, although chairman Edwards did not seem to be unduly worried and attended his farewell dinner from the Hardie board later that evening. Cooper consoled himself with the thought that at least he had placed the company on notice that something was wrong.

Foundation complaints to Hardie about its finances during ensuing months were met with similar brushoffs. One exception to this pattern of apparent indifference was a near-emergency triggered by a successful asbestos claim against Hardie's former brake subsidiary that placed it perilously close to insolvency. Alarm bells rang at Hardie, which was about to seek Supreme Court approval to move offshore. The bankruptcy so soon after one of the former Hardie subsidiaries had become part of the foundation would not help Hardie in court. Shafron was worried that the court's public proceedings might offer critics a chance to object: '... stakeholders could be expected to argue that funds in the foundation are insufficient to meet all future claims, that JHIL has a legal liability to future claimants, and that if approved, the scheme will mean that JHIL is left in a position unable to meet that liability ...'[61]

The Hardie board quickly solved the problem by accelerating payment of $1 million under the indemnity deed that ensured the parent would not be sued for injuries caused by its former brake subsidiary.

To obtain court approval for his planned share swap between the original Australian parent and its new Netherlands identity, Shafron recommended that the Hardie directors adopt the course he thought had the most 'cosmetic' appeal.[62] The new Dutch company (JHI NV) would buy shares in the Australian James Hardie Industries Limited in exchange for its assets. The

Australian company would then reduce its capital by an equivalent amount, paying a substantial dividend to its Dutch parent. This process would be conditional on the Dutch company subscribing to partly paid shares in its Australian predecessor, so that at any time the former JHIL could demand that the shares were paid in full, giving it access to capital up to the value of its previous assets. Even though the former Australian parent would be left as a shell company, with '… what would be judged by the future JHIL as an altogether unnecessary capital lifeline', Shafron argued the scheme would disarm objectors who could delay the scheme's approval: 'If JHIL is left in the same economic position after the restructure as it was in before, then stakeholders should effectively be deprived of grounds for complaint.'[63]

By July 2001 the Hardie board had signed off on Shafron's strategy and applied for a Scheme of Arrangement with the NSW Supreme Court. Despite Shafron's concern about spoiler actions, there were none. What scrutiny there was came from the court.

HARDIE GOES DUTCH, proclaimed a headline as media attention focused on the tax advantages for Hardie.[64] Commentators predicted other companies would follow Hardie's example because of Australia's unwieldy tax laws. None mentioned the well-advanced negotiations between the US and Australian governments to reduce withholding tax to zero. Nor did anyone mention the implications for Hardie's recently separated subsidiaries and their asbestos claimants.

Asbestos claims, however, *were* specifically raised in a list of questions from the Supreme Court's Justice Kim Santow following Hardie's application.

'What effect will this [the proposed scheme], if implemented, have on asbestos claims against James Hardie?' he asked.[65]

The Allens lawyers wrote back:

... the Scheme will not affect the position regarding asbestos claims ... JHIL has at times been joined as a party to such proceedings, but has always successfully resisted any claims against it ... That said, it cannot be said that JHIL will never be held liable. JHIL will have, through existing reserves and access to funding in the form of the partly paid shares, the means to meet liabilities which will or may arise in the future whether in relation to asbestos-related claims or other obligations to other persons.[66]

Despite a warning from one of the Allens lawyers, David Robb, Hardie failed to inform Justice Santow about the indemnity deed against asbestos suits Hardie had quietly secured from its former subsidiaries. Robb later agreed that this omission was to avoid 'a rigorous investigation' of the transactions between Hardie, the foundation and the subsidiaries.[67]

Shafron also rejected Robb's advice that the court might need to see the Trowbridge studies: 'My point on T [Trowbridge] is really code for 'the thing is not that defensible'!'[68]

Justice Santow paid particular attention to what he called the 'fundamental matter' of the partly paid shares the Netherlands company JHI NV would hold in the former Australian parent, asking whether there was '... any possible basis upon which a call upon partly paid shares upon a Dutch company could be resisted under Dutch law?'[69]

Hardie's replies were unequivocal. The purpose of the shares was '... to ensure that [JHIL] has access to funds going forward to meet any potential liabilities'. JHIL would be able to call 'any or all' of the remainder of their issue price '... at any time in the future and from time to time'.[70]

Just what Hardie meant by 'any time in the future' was not revealed, but Justice Santow was certainly persistent on the point:

SANTOW: When it says JHIL will be entitled to call upon JHI NV in the future and from time to time, is that right?
HARDIE: Yes.
SANTOW: There is no time period laid down?
HARDIE: No.[71]

The Justice reinforced his concern: 'One would need to make sure every step is taken ... to ensure that a call must be met. In other words, there must be the clearest possible Dutch exchange approvals required if it is possible to get them in advance in order to ensure there is no blockage in the flow of funds to Australia.'[72]

In a Supreme Court hearing of this nature, where no opposing parties appeared, the onus was on the applicant to present all material facts. Robb later admitted that Hardie had always considered the option of cancelling the partly paid shares, but Santow was never informed. Justice Santow expressly asked the Hardie lawyers if there were any other matters it should tell him about. They could think of none.[73]

Outside the court a storm was brewing. Unknown to Santow, the foundation's predicament was getting worse. Its directors had commissioned a new Trowbridge study, which predicted its future asbestos liabilities would be $574 million. The outlook for its earnings was equally bleak: the return was more likely to be 8.7 per cent rather than Hardie's prediction of 11.7 per cent.[74]

Chairman Sir Llew Edwards warned Hardie's CEO Peter Macdonald that the foundation had a 'very limited life', possibly less than ten years, adding that he would write a formal letter to Macdonald setting out the predicament. The letter, dated 24 September 2001, stated that the foundation would need an additional $200 million in order to last twenty years.[75] Macdonald later claimed he did not even read the letter until mid to late October.[76] The dates were significant because four days after the foundation's letter was sent, Hardie sought shareholder approval for its move offshore before returning to Justice Santow for his final permission. Neither the court nor the shareholders were told the foundation had placed Hardie on notice that it required a cash infusion to avoid bankruptcy.

Ironically, only hours before the shareholders meeting at Sydney's Regent Hotel, Hardie's chairman, Alan McGregor, informed his fellow directors that the main reason for the corporate relocation to the Netherlands as stated in the company's information memorandum was about to disappear. The Australian and US governments had just announced an in-principle agreement to abolish withholding tax on dividends paid from US earnings. There would no longer be any tax advantage for Hardie to move to the Netherlands.

McGregor assured his board that the tax change would be unlikely to take effect for several years.[77] It pressed ahead. Two hours later the assembled shareholders voted overwhelmingly in favour of the scheme and the following week the NSW Supreme Court gave its approval. The shareholders now owned shares in

James Hardie Industries Netherlands, which ran the business of the group. JHIL had morphed into JHI NV.

Shafron joked with Macdonald that they should answer the foundation's letter about its dire financial straits in 'four years'.[78] He drafted a reply rejecting any inference that the funding shortfall was Hardie's fault. Macdonald did not send his written response for a year. Hardie's board similarly unconcerned, instructing its CEO to continue to 'manage' the situation as he was doing.[79]

The foundation directors remained eager to coax additional funds from Hardie NV. When Macdonald suggested they withdraw the letter to avoid an unproductive 'pissing contest', they decided to enter into 'constructive dialogue'.[80] They met the CEO and Shafron early in 2003 at the Qantas Chairman's Lounge at Sydney Airport, where they pressed their case, claiming that the foundation's predicted life before bankruptcy had now shrunk to an alarming five or six years.

Macdonald was unmoved, shielding his eyes when Cooper brandished the gloomy projections and protesting, 'We can't get too close, don't need to know and can't get into the detail.'[81]

But Hardie NV still needed the cooperation of the foundation directors for a final manoeuvre. When the foundation was created, an agreement was included for Hardie NV to put the shell of its former parent, James Hardie Industries Limited (now renamed the eminently forgettable ABN 60) under the foundation's control. Macdonald suggested that he might release an extra $10 million to the foundation if Hardie exercised that option. He told the directors that he could even accelerate the periodic payments for the indemnity now that Hardie was cashed up after selling gypsum interests in the US.

What the foundation's directors did not know was that Hardie had also begun secret preparations to cut the $1.9 billion lifeline of partly paid shares, now the last remaining link between its new Netherlands parent and its asbestos past in Australia.

Allens' lawyers met to consider whether it was too soon to cut the lifeline following Hardie's appearance before the Supreme Court, and asked themselves the question: 'If [we] had this in mind last Oct[ober], should [we] have disclosed?'[82]

The crucial question of timing was again considered a fortnight later, as noted on the law firm's files:

- This is all very soon after having been to the market with a scheme booklet which did not state that reduction of the partly paid shares was part of the intentions for JHIL ...
- ASIC may also take an interest and choose to investigate
- Liquidation in particular, so soon after the scheme may raise too many questions ...[83]

Further discussions between the lawyers canvassed what documents might be revealed under a court challenge, and what evidence those privy to the scheme's planning might give:

Whatever misleading conduct that may be alleged has already occurred ...

Critical q., at time of scheme, we had intention to cancel these partly paid shares? ...

If had int[ention] cancel pp shares, or t/f [transfer] pursuant to the put option then misleading not to include in scheme docs[84]

As their deliberations drew to a close, the lawyers wondered whether the company would need to disclose publicly its plan to put ABN 60 under the foundation and cancel the shares.

> Legally risk may not be higher
>> Commercially/Politically of higher risk
>> Nothing Santow can do.
>> ASIC might do something ...
>> Say, no int[ention] to t/f [transfer] at time of scheme.
> Didn't cross anybody's mind to do this. Reason had partly
> paid shares was to have greater flexibility. Had an intention
> to deal with it later.[85]

What exactly did 'Say, no intention ...' mean? The Allens' note-taker later explained that he had used the word 'say' in the sense of 'assume'.[86]

Shafron met the foundation directors again in early July, reporting to his colleagues afterwards that they were 'not suspicious' about taking custody of ABN 60. He had mentioned the partly paid shares, 'but don't think they appreciate it'.[87] The Hardie NV executives were nonetheless becoming frustrated by the company's lingering asbestos legacy, which proved to be 'an unwelcome distraction' to investors when its debt was renegotiated.[88]

But at least one foundation director was in no mood to negotiate. The following week, Peter Jollie, the only board member who had no previous association with Hardie, made his frustration clear in a tetchy exchange. The foundation needed $200 million, he told Macdonald, dismissing the $91 million offered for ABN 60 and the $15 million sweetener as a 'drop in the ocean'. Hardie could provide the money if it 'felt it was important enough'. He warned of a difficult time ahead '... particularly from a PR perspective'.[89]

Meanwhile, Dennis Cooper approached NSW Premier Carr at a Labor Party election fundraiser to brief him on the foundation's plight. Carr scribbled some notes. 'I just told him the facts, as simply as possible,' said Cooper. 'We were running out of money.'[90]

He also forwarded the foundation's figures to the NSW Cabinet Office. The government was now on notice that trouble was brewing.

By early 2003 the funding shortfall was forecast to be $500 million. Even Cooper's usual stoicism was wearing thin when he reported an exchange with Shafron to his fellow foundation directors.

'They say they cannot meet requirements for additional funds; would 'blow apart' the separation. They again 'defended' the extent of assets originally provided, despite non-use of latest data. They confirmed that they are concerned about the PR risk to them but, again, say they cannot do anything.'[91]

The mood of the foundation's directors was not improved when they saw a smug-looking photograph of Macdonald in the *Australian Financial Review* announcing that Hardie was about to make a $250 million cash issue to its shareholders.[92] Since its gypsum sale, money was evidently not a problem for James Hardie Industries NV. 'If we don't see a need for it,' Macdonald told the newspaper journalist, '... we can give it back to [shareholders].' There was no hint the company might give more to the foundation for asbestos victims.

One of the directors decided it was time to quit. Michael Gill knew more than his colleagues, having previously advised

Hardie on the restructure. Citing the demands of his law firm as the reason, he resigned from the board, while maintaining his link as a consultant.

Gill's replacement, Ian Hutchinson, was soon to profoundly change the foundation's approach to its funding crisis and its attitude to Hardie. The foundation was to make it clear to JHI NV that it would not be signing any deals until more money was forthcoming.

But Hardie was unfazed. It had determined to sever the link and that was what it would do. Chairman McGregor and his CEO agreed to persist with their plans regardless of the foundation's resistance. Macdonald briskly instructed his team: 'No communication with the foundation for now.'[93]

Hardie moved swiftly and silently to accomplish its final manoeuvre, one Macdonald wrote would '... make even clearer the separation of JHI NV from any asbestos legacy'.[94] If Hardie could shed its old parent before 31 March 2003, it could claim significant tax benefits. Lacking a pliable foundation, Macdonald simply formed a new company called ABN 60 Foundation into which the shell of the parent and Hardie's former asbestos subsidiary in New Zealand would be placed.

A board committee of McGregor, Hellicar and Macdonald convened. The foundation had already placed them on notice about its precarious financial situation and knew the implications for people yet to fall sick from Hardie's asbestos. It had told the company that within five years it would have no more money and 80 per cent of asbestos claimants would receive no compensation.

Yet the board committee approved the cancellation of the partly paid shares, thus removing the last remaining lifeline to Hardie's operating business.

The entire operation took less than a month and was executed in perfect secrecy. There was no interest expressed by outside observers, whose attention was gripped by the US invasion of Iraq launched that week.

Macdonald and Baxter conferred: no disclosure was necessary, they decided.[95] Neither the public, the shareholders, the court, nor the foundation needed to be told.

eleven

BEHIND THE SCENES: ASBESTOS POLITICS

Election fever gripped New South Wales in the early months of 2003 as Bob Carr campaigned for a third term as premier. Despite being briefed on the plight of the foundation which housed Hardie's former asbestos companies, he had given no indication of what his government might do if asbestos claimants were stranded without compensation.

James Hardie executives hoped there was a good chance he might change the law and wind back the unique procedures of the Dust Diseases Tribunal. So did Sir Llew Edwards, the foundation's chairman, who, like his former Hardie colleagues, was convinced the tribunal was out of control. Edwards had lobbied for a new compensation regime since taking the job and tipped that the Carr government would introduce one if it won the election.[1]

But Edwards' near-bankrupt foundation was also changing. With the departure of Michael Gill from its board, his replacement, Ian Hutchinson, was to have a profound effect on its behaviour. Hutchinson, an experienced lawyer, had been Lloyd's Australian representative when the reinsurer had almost collapsed under the weight of asbestos claims. Initially, both Edwards and Hardie managers welcomed his recruitment. They soon changed their minds.

Peter Jollie was already resisting Hardie's offer of extra cash in exchange for an indemnity against legal suits. Now Jollie was joined by a fellow dissenter in Hutchinson, who at his first board meeting was aghast to realise the other directors were hovering on the brink of accepting Hardie's deal.

'Nobody had told me about this. When I dug into this deed, thank God for Jollie, because he stood up and he wasn't going to agree to it. Not for another $20 million when they are $500–600 million short. I stood up and said 'no way!'"[2]

Hutchinson had laid down two conditions before he accepted a job on the board. Because the foundation's plight 'fails the smell test terribly', he demanded the right to choose a lawyer to 'turn the thing upside down and inside out'. Second, he wanted to nominate someone to sort out the foundation's insurances, which he later described as a 'total shambles'.[3]

The lawyer Hutchinson chose was Nancy Milne from Clayton Utz, a tenacious solicitor who quickly discovered the foundation had very few records about the circumstances surrounding its own launch. She demanded them from Hardie's lawyers at Allens, but hit a brick wall.

Within a few months the foundation's relationship with Hardie transformed from that of a supplicant begging for more funds to a potential litigant. The NSW government, too, was to hear less from the foundation about the need for reform of the Dust Diseases Tribunal; it was urged, rather, to investigate Hardie's behaviour. It was a dramatic change in direction for the foundation.

Even with its asbestos subsidiaries separated, Hardie's board had never lost its desire to see the NSW compensation system wound back. Lacking the credibility to lobby for such change itself, it had hoped the foundation, with its veneer of concern for medical research, might prove a more effective vehicle to pressure the NSW government.[4]

Edwards embraced this cause with crusading zeal, regularly providing progress reports on his lobbying efforts to his former colleagues.[5] The more parlous the foundation's finances became, the more he saw need for change. He persisted in this advocacy to me even in 2004, when the foundation was close to liquidation.

The problem was that the NSW government would never remove the DDT and make asbestos liabilities a compensable disease through workers' compensation, as happens everywhere else. I put that to government on a number of occasions. If the government had brought in a scheme ... if we'd had underwriting, we could have lasted another three or four years ... The only way this'll be resolved is with some form of statutory scheme, probably capped.[6]

Edwards, a former politician from Sir Joh Bjelke-Petersen's conservative Queensland government, told me he had been encouraged by his talks with NSW ministers. Indeed, he claimed that one 'very senior minister' had told him, 'Your bloody Liberal friends caused this problem by establishing the DDT and then the stupid Labor government put all their mates on it!'[7]

Premier Carr had certainly shown enthusiasm for shaking up the state's compensation regimes. Soon after the foundation's launch, Carr delightedly taunted a picket of angry unionists outside Parliament House with a victory sign before enacting legislation that capped workers' compensation payouts. He had declared war on what he called US-style 'litigation culture' and 'Santa Claus' payouts.

Carr had found a ready ally in the Insurance Council of Australia. A leaked letter written to the government after the premier and his treasurer attended the council's NSW conference in March 2002 revealed that the asbestos regime was also firmly in the industry's sights: 'I refer to our meeting at the ICA dinner ... and in particular our discussion in relation to asbestos disease issues and the court system ... You indicated a desire to look further into this issue and suggested I put pen to paper to address points of concern.'[8]

The attached brief echoed changes Hardie had long wished for, recommending the wholesale scrapping of the 'plaintiff-friendly' tribunal and replacing the right to sue under common law with a no-fault capped statutory scheme. Carr did not remember receiving the letter. If he did, he told me later, it made no impression. His focus was on the much bigger issue of liability insurance, he claimed, adding he would have simply passed the letter on to the bureaucracy.[9]

Hardie and its advisers monitored these developments closely, noting a call by a federal minister, Joe Hockey, for a National Accident Compensation regime similar to New Zealand's. CEO Macdonald reported to Hardie's board: 'Premier Carr in NSW has strongly supported the proposal, saying we must stop a trend towards a culture of litigation ... James Hardie would support such legislation and is looking at how it can best assist the process of getting legislation in place.'[10]

Macdonald also wrote that Carr's minister John Della Bosca, who had spearheaded the changes to motor vehicle accident and workers' compensation laws, was unhappy with the costs of the

Dust Diseases Tribunal and was 'prepared to take on the DDT with similar reforms'.

Hardie had bequeathed the foundation another weapon in its armoury to convince Carr to wind back the tribunal. Allens had launched a test case against the government over Warren Hay, the fitter who had worked in a state-owned power station and later developed mesothelioma.[11]

The case had become a marathon legal battle, heading for the High Court after an initial loss for Hardie's former asbestos subsidiary before the tribunal. A win could mean hundreds of countersuits against the NSW government for failing to exercise a duty of care to protect workers through its health inspectors. The foundation was holding Hay's case as a gun at Carr's head, one that Sir Llew Edwards was not frightened to mention.

Macdonald summarised a briefing about the case from Edwards: 'Although the foundation thinks this may be a chance to sue Governments, it is most likely going to be a form of leverage on Governments to act responsibly in managing current asbestos costs ... The NSW Government is quite highly motivated to do something about the issue — Sir Llew has been interacting with Carr's department and the role and continued existence of the DDT is being questioned.'[12]

As the 2003 election approached lobbying intensified. The foundation's actuaries at Trowbridge suggested they do a scoping study on asbestos liabilities and the cost of an alternative compensation scheme. They offered a 'Proposal for Stakeholder Management Assistance' to build relations with asbestos defendants, insurers, lobby groups and government officials 'sympathetic to liability reforms'.[13] The giant insurer Allianz also cranked up its campaign against the tribunal. Allianz was already waging a similar battle across the globe, where the insurance industry had rekindled a controversial bill in the US Senate to cap asbestos claims.

Carr was re-elected and within a few weeks the NSW Cabinet Office convened a meeting to discuss possible changes to asbestos compensation. The Cabinet Office director-general, Roger Wilkins, was later vague in his recollection of the subcommittee established to address the issue.[14] Kate Mackenzie, Wilkins' former deputy before heading Della Bosca's department, participated in the discussions. She had chaired the Dust Diseases Board meeting which approved Hardie's foundation when it was first launched.

Amendments to the tribunal legislation were drawn up for the Cabinet Office. Caucus was warned to expect a briefing. The bureaucracy buzzed with talk of a 'third wave of reform'. But Allianz overplayed its hand. Eager to enlist 'stakeholder' support, it arranged briefings for individual trade unions. Word about the impending legislation soon leaked out.

An outraged NSW secretary of the AMWU, Paul Bastian, fired off angry letters to government ministers, telling journalists: 'Bob Carr has a lot to answer for in this state for the benefits he ripped out of workers in his last round of workers' compensation. And Bob Carr is going to have a lot more to answer for if he's contemplating ripping the benefits out of the victims who die from asbestos-related diseases.'[15]

Allianz was caught off-guard by the attack. Under media pressure, the insurer issued a statement welcoming any changes that would both reduce 'unnecessarily high legal fees' and interstate claims from people who have '... never set foot in NSW, let alone worked here'.[16]

Protests flooded in from interstate asbestos support groups deploring the inequity of excluding former NSW residents who had moved interstate to retire only to discover their twilight years were to be cut short by mesothelioma.[17]

The beleaguered attorney-general, Bob Debus, confirmed to the media that legislation to stop interstate tribunal claims was under consideration, but any changes would be 'minimal'.[18]

Debus claimed later he was as surprised as anyone by the Cabinet Office initiative.[19] He had already warned his colleagues the previous year that the political sensitivities associated with asbestos made changes unwise. The attorney-general dispatched his staff to collect data from the tribunal. He soon discovered that the percentage of interstate claims was tiny. Almost all came from people who had either lived in NSW at the time of their exposure to asbestos, or who had used asbestos manufactured in NSW.[20]

Hardie's former companies, it emerged, were then listed as a defendant in 44.7 per cent of tribunal cases. Debus found that the overwhelming majority of claims before the tribunal were settled before hearing and almost all of the ones that were not resulted in a verdict for the claimant. The average award for a mesothelioma was between $150,000 and $180,000. The tribunal, he decided, was working as it should.

Debus used these figures to argue in Cabinet against the changes. He also sought agreement from his interstate counterparts for a contribution to the administrative costs of claims from their states. For now, at least, a decision was deferred.

The pressure on the NSW government remained intense. A week later, the foundation won in the High Court with the marathon Hay case, which now returned to the Court of

Appeal. Another win would mean the government faced a vastly increased asbestos liability of its own. The case was listed to be heard in April 2004.

But the foundation, under the influence of its new director Ian Hutchinson, was not winning its battle to extract information from Hardie. Hutchinson's hand-picked lawyer, Nancy Milne, briefed Michael Slattery, a Queen's Counsel with a reputation for fighting cases of principle. The two drafted an extensive letter to Hardie summarising the chain of events leading to the foundation's lack of funds and demanding all relevant documents as part of pre-trial discovery. They copied the letter to Attorney-General Debus.

According to Milne, the reply from Hardie's lawyers at Allens was clear: 'We were basically told to go jump. They sent us back several folders of documents which were full of irrelevant material. They told us that we were wasting our time.'[21]

During subsequent weeks the foundation directors and lawyers met with ministers Debus and Della Bosca and briefed staff in the premier's department. No longer was their focus on compensation reform, but on the need for legal action against Hardie, whom they accused of deliberately abandoning its underfunded trust. They wanted an inquiry.

Debus and Della Bosca had initially treated the foundation's approaches with some scepticism, but they were now starting to believe its story. While Della Bosca had hardly endeared himself to the unions with his tough stand over workers' compensation, he was aware that asbestos disease was an iconic issue for the labour movement.[22]

Hardie's public profile at this time could hardly have provided a starker contrast to its private dealings. A fortnight after Allens had rebuffed the foundation's attempt to obtain information, Hardie's CEO Macdonald wrote an op-ed article in the *Australian Financial Review* urging Australian companies to embrace stock exchange guidelines for corporate governance:

> The message is clear. Corporate Australia must respond to the legitimate concerns of disaffected investors, restore its credibility and recapture the trust and confidence of shareholders and the wider community ... Companies should view the guidelines as an opportunity to undertake highly self-critical self-assessment to test whether their corporate governance model has integrity and provides a sound moral framework for ethical decision-making.[23]

Macdonald's PR chief, Greg Baxter, had recently won the company a string of awards for corporate disclosure.[24] Even Hardie's former chairman John Reid, who made a rare public appearance to join the chorus against 'US-style litigation',[25] had just published a book, *Commonsense Corporate Governance*, hailed by the Australian Institute of Company Directors as a courageous moral blueprint for aspiring board members.[26]

Three weeks after his public advocacy for corporate governance, Macdonald took the company's case directly to Debus and the Cabinet Office, stressing that the foundation's cost blowout was proof of the urgent need for a change to the tribunal. He met the foundation directors the following day, and told them Debus had said 'reform' was 'definitely' on the agenda.

> The government stated that the insurance industry had been 'most unhelpful' with the publicity surrounding their

submissions. They forced the government to delay any response to reform issues. The government said that public campaigning for reform is not the way to achieve results. It is better for interested parties to work in the background.[27]

The foundation directors were unimpressed. They still needed more money and believed they had been misled. 'Let me be totally clear on this,' the minutes record Macdonald as saying. 'The company does not contemplate making any offer whatsoever and no settlement is possible.'[28]

Three weeks later the foundation went public with a media release, although even now its message was muted. The key point was buried seven paragraphs in, on the second page: Hardie had told the foundation there would be no money and its funds could be exhausted within five years.[29]

For the first time the public had been told: the foundation had been underfunded and was heading for bankruptcy. The following day both Hardie and the government responded.

Macdonald expressed 'surprise' at the 'dramatic departure' from previous experience of asbestos claims, adding that major groups within Australian society were looking at '… reform of this area of litigation'.[30] Minister Della Bosca was by now not looking at reform, but at Hardie itself, stating the government was examining every option '… to pursue all avenues against the company'.[31]

Della Bosca believed only some kind of Royal Commission would be equipped to find out what had happened, he later recalled.

'We just had to make sure the Royal Commission had

sufficient powers to get whatever evidence it could of what the skulduggery was and how it was done.'[32]

But the foundation heard a very different message when it visited the Cabinet Office, where Director-General Wilkins preferred a parliamentary inquiry, as he later admitted.

'I had cautioned Carr about setting up a Royal Commission and told him to think about using a parliamentary inquiry to flush out documents and air issues.'[33]

Wilkins told me he was concerned about the cost and feared that such an inquiry could 'go all over the place', in particular, questioning '... what the health department and the WorkCover inspectors had been doing all those years'. This was the fear that Hardie's test case over Warren Hay had been designed to provoke. Ironically, Hardie's Greg Baxter had earlier contemplated the government might set up such a restricted inquiry to show it was 'doing something', which he regarded as 'a relatively meaningless' option.[34]

In the event, the push by Debus, Della Bosca and the foundation for a Special Commission was decisive and Wilkins was overruled.

Peter Macdonald soon found he was unable to meet with government ministers. Publicly, he continued to proclaim that Hardie's behaviour had been beyond reproach. In an interview about the foundation's plight I recorded with him for the ABC's *7.30 Report* (returning to the Hardie saga after a gap of almost twenty-five years) he stuck to the company script:

MACDONALD: I believe that James Hardie has done the right
thing, not only the legally correct thing, but also the morally
right thing, in attempting to fully fund those obligations ...

PEACOCK: You're saying that James Hardie as it exists now in the Netherlands has no legal liability?

MACDONALD: James Hardie has always done the right thing in relation to any company it's involved with ...[35]

It was like talking to a friendly robot. In a few short sentences he mentioned 'right/proper/moral thing' and 'fully funded' twenty-two times.

Privately, the Hardie board was growing concerned about the direction the NSW government might be taking. Director Meredith Hellicar tried to help the Hardie management team arrange a meeting.

'I personally became sufficiently frustrated and said, 'Come on, they *can't* not be seeing you!' I rang up someone I knew to try to get in: 'Say it's the director of the company ringing up saying this is ridiculous.'[36]

But the government had made up its mind. Carr had bowed to the arguments of his ministers and agreed to a Special Commission. Debus also overruled a Cabinet Office nomination for commissioner, preferring instead the former Federal Court judge David Jackson, QC, a choice he claimed Director General Wilkins opposed.

'He did all he could to stop it. There was a long period — several weeks — where he actively resisted the appointment, but I insisted on it. I knew that Jackson's prestige would have a massive impact, as one of the two best High Court–level barristers in Australia. Wilkins was adamantly opposed.'[37]

Wilkins later told me he hadn't expressed a view either way.

Hardie soon learnt of the planned inquiry, although Hellicar claimed the company was privately reassured by the government that it still wanted to change the compensation system: 'But the message we got back from various advisers ... was that the government said we are not going to be able to reform the system until we have shown that everything that James Hardie did was OK.'[38]

In early February, just days before Cabinet and the governor were due to sign off on the inquiry, Macdonald called Della Bosca from California; the minister agreed to meet with him. Della Bosca later shook his head as he described the meeting:

> He wasn't angry, he wasn't rude, he just very calmly said to me: 'You can't do this to us. You're a pissy little provincial government. You can't stop us. We're now a global company and we have done what we think was in our shareholders' interests. That's my job and that's what I've done. And if you have a different view, well, you go get yourself a multi-billion-dollar company, become its chief executive and you can have a different view.'[39]

From the outset Hardie's CEO approached the commission hearings with the same hard line. The company had no legal liability, Macdonald told a disbelieving media. The foundation had been provided with nearly $300 million, '... which is what experts said at the time was needed to provide certainty for future claimants'. He told ABC Radio he could not understand how there could now be a funding shortfall: 'This is something which is really surprising. Is there a problem? What has driven a threefold increase in costs in just three years? What could it be? Could it be legal costs? Other system costs?'[40]

Hardie's subplot was still clear: it was time to change the compensation system. To reinforce this point, Macdonald claimed that the James Hardie Group was only responsible for '... around 15 per cent of the asbestos disease in Australia'. In fact, it later emerged that by 2002 James Hardie companies had been sued in 48 per cent of claims before the DDT, and it was the sole defender in 35 per cent of cases.[41]

In the weeks preceding Commissioner Jackson's inquiry, two competing asbestos plaintiff law firms, Turner Freeman and Slater & Gordon, joined forces to represent unions and asbestos victims groups, briefing Jack Rush, QC, as their counsel. Rush and the foundation's silk, Michael Slattery, were already friends through their membership in the Naval Reserve. Another unexpected party joined the alliance against Hardie.

The foundation's lawyer, Nancy Milne, received a phone call from CSR's David Miller.

'He said, 'There is a coalition of common interests here. Why don't I organise for you and the unions to get together?" recalled Milne with a smile. 'That was the beginning of a sort of partnership.'[42]

CSR had been deeply suspicious of the Hardie restructure from the previous year, when an executive had warned Macdonald of his concern that CSR would be 'holding the baby', as the last asbestos defendant left with the deepest pockets, if the foundation ever ran out of money.[43] Macdonald had indignantly explained that the lifeline of $1.9 billion would prevent that. Although his response had done little to mollify CSR, it had yet to realise that the partly paid shares had been secretly cancelled.

* * *

The foundation agreed to provide its evidence first. In the scramble to collect its directors' statements, Milne was told by a colleague sent to interview Michael Gill, who had resigned earlier that year, that '... all he could do was look at his feet'.[44] Gill soon called to tell Milne she could not represent him. Milne immediately suspected that he might have earlier advised Hardie on its restructure.[45] She was right.

In April the Special Commission convened. A courtroom at the tribunal had been set aside for its hearings. Such was the public interest that Jackson permitted TV cameras to film the first few minutes of proceedings as an army of silks introduced themselves. Bernie Banton made a point of attending the hearings every day he could, sitting behind the lawyers where '... I can see absolutely eye-to-eye with Commissioner Jackson'.[46]

In the second week when Gill gave evidence the unions and asbestos groups smelt a rat. Why did he need his own lawyer? They persuaded the Counsel Assisting the Commission, John Sheahan, SC, to subpoena his documents, and the answer became clear. Back came notes of Gill's earlier advice to Hardie on achieving an 'unbridgeable corporate split' with its asbestos companies. He had told nobody — not his fellow directors, not the commission.

Gill explained that he had not thought the information was relevant because his advice to Hardie had been given six months before the foundation's creation and involved meetings lasting only about two and a half hours.[47]

Milne was appalled: 'Michael was basically sitting there controlling the process and none of his fellow directors knew that he had previously advised Hardie. Not to tell people and not to 'fess up to his fellow directors was really extraordinary.'[48]

When he was recalled to give evidence a second time Gill failed to appear. Although the reason he gave remained confidential, it was believed to be on medical grounds. In any case, he later returned to work at Phillips Fox.

Hardie's lawyers at Allens continued to make life hard for the other side. 'The Hardie documents came in dribs and drabs,' recalled Milne. 'I remember having screaming matches on the phone with Allens about that, because they made it as difficult as possible.'[49] When the company's key witnesses, Macdonald and its general counsel Peter Shafron, were due to give evidence, their documents were only made available the afternoon before. Paralegals worked through the night scrambling to copy, collate and index them.

For the asbestos disease sufferers and their families the appearance of the Hardie executives in the stand was the moment they had been waiting for. On the day Shafron was sworn in Banton was joined in the courtroom by others from the Asbestos Diseases Foundation of Australia, among them the newly widowed Elizabeth Thurbon. She and the other women wore black hats in the courtroom as a reminder of their dead husbands.

'I watched him intently as he gave evidence. He deliberately turned his head away from me and the others in black hats as he just couldn't bear to look at us. His main reply to every question was 'I can't recall'.'[50]

Shafron's fuzzy memory also irritated Commissioner Jackson, who curtly instructed him not to respond '... in a way that makes it apparent that you're not trying very hard'.[51] Under cross-examination from the counsel assisting about his refusal to give the foundation directors the latest actuarial study of Hardie's asbestos liabilities because it was 'not complete', Shafron conceded he had misled them.[52]

When Hardie's CEO Macdonald gave evidence the following week, the tide had begun to turn against the company. Cool and combative, Macdonald quickly earned the media sobriquet 'Rocky' for his verbal sparring with commission lawyers.[53]

It reached absurd lengths when he was quizzed over Greg Baxter's decision to blame the lack of a live webcast of the foundation launch on a 'technical difficulty':

Q: ... you understood that he was proposing just a little white lie on the website, correct?

MACDONALD: He's saying he could explain it, yes, on a basis of technical difficulty, I agree.

Q: Dishonestly?

MACDONALD: Well it's a proposal, it hasn't actually been carried out yet, in fact I need to —

Q: The proposal is dishonest?

MACDONALD: If Mr Baxter used that explanation then it would be not correct ...

Q: And to make an incorrect statement with knowledge of its falsity would be dishonest, would it not?

MACDONALD: If that's the definition of dishonesty then that would be the case.

Q: Does it work for you?

MACDONALD: Well, you're the linguistics expert ...[54]

The CEO became a target for media ridicule, eluding protests from Banton and others via the commission's fire escape and security entrance, and during lunch breaks bolting to a small interview room, its tiny window taped over.

Despite the growing anger against Hardie in the general community, its critics at the inquiry were becoming increasingly apprehensive about whether it could ever be called to account. Concern heightened when they learnt midway through the proceedings that Hardie NV had cancelled its partly paid shares to the former Australian parent. CSR, too, was 'very, very angry', according to Turner Freeman's Armando Gardiman.

Their attention was now firmly focused on the commission's final term of reference, which asked whether any reform was needed to the Corporations Act for the foundation to manage its asbestos payments. It certainly looked like some change to the law might be needed to claw back Hardie's money from the Netherlands, one of the few industrialised nations which had no treaty with Australia to enforce court judgments. Peter Gordon, senior partner of Slater & Gordon, asked the secretary of the Australian Council of Trade Unions (ACTU), Greg Combet, to become more involved with the commission;[55] Combet soon began to attend the hearings in person. And he also began talking privately to Premier Carr about what could be done once Jackson reported.[56]

James Hardie's carefully controlled media image was shattering. The tabloids bayed for blood. Bernie Banton had been discovered as good media 'talent'; scarcely a day passed without his increasingly familiar face with its trademark oxygen tubes appearing on TV news bulletins. Inside the company, matters appeared to be lurching out of control. Greg Baxter, the PR chief, predicted at the start of the inquiry 'it was obviously going to be very ugly'.[57]

About a week into the hearings Shafron called from California, asking, 'What the fuck's going on with the press reporting?'

'You're joking!' Baxter answered. 'It'll be worse in two months!'[58]

For Baxter no media scenario was too silly. It was no longer his problem. With exquisite timing, he had negotiated a new job with Rupert Murdoch's News Corporation, quitting James Hardie just before Macdonald gave evidence. He said later he 'felt sorry' for his colleagues. There was a rapid turnover of Baxter's replacements. His assistant soon took maternity leave, never to return. Her replacements had their work cut out catching up on the complex corporate history and dealing with an overtly hostile media.

One colleague, John Noble, strongly recommended media contingency planning, but was ignored as the Allens lawyers insisted the company maintain its legal defences. Noble, too, left the building one day never to return. Suffering from severe depression which his colleagues said was intensified by the Hardie experience, he committed suicide a few months later.

Quietly, one controversial Hardie adviser had returned. Stephen Loosley, the former ALP senator and NSW secretary who had lobbied against government intervention when the foundation was launched, continued to advise the company. As an ex-senator, a company insider told me, Loosley had extra credibility with the company's US-based directors, who mistakenly equated the position with that of a vastly more powerful US senator.

* * *

As Jackson's inquiry drew near its end the most sensational evidence was yet to be given. The conduct of Hardie's lawyers themselves had come under scrutiny and Allens soon engaged a counsel to represent the firm. Commissioner Jackson expressed surprise that this had not occurred sooner. The commissioner had already issued a blunt warning: '... why didn't someone just stand back and say "This is just too hot", or words to that effect, and a view might be taken is that the appropriate people to say that would have been their [Hardie's] solicitors.'[59]

Late in the afternoon on 11 June, the senior Allens lawyer David Robb was quizzed about the company's failure to inform the NSW Supreme Court that the partly paid shares might be cancelled. Had he checked, he was asked, if the company had then intended to cancel them?

Robb took a breath, and then answered: 'I did consider whether or not the court had been misled and in response to this question ... I drafted ... advice on this topic that ... indicated that the court may have been misled and also the information memorandum may have been incorrect.'[60]

The courtroom was silent, apart from the furious scratch of journalists' pens and the hiss of Banton's oxygen tank. All eyes were on the Allens solicitor.

Robb continued: 'After I drafted that advice I went on five weeks' holidays. The matter was transferred to Richard Alcock and Michael Ball — who was already acting — continuing to act. On my return, that letter had advanced and had been sent out.'

It was the most dramatic moment of the commission hearings.

Robb appeared to be choking back tears. He had made no previous reference to this advice in his earlier evidence or statements, nor did any documents suggest it might exist.

The barrister for the unions and asbestos groups immediately demanded that Allens supply his draft and any records associated with it. When the missing evidence arrived it revealed that during Robb's holiday Allens senior partner Michael Ball had deleted the explosive paragraphs. The advice Allens then sent made no reference to the possibility that the Supreme Court had been misled.

Commissioner Jackson invited Ball to give evidence. He refused.

Banton told me later he had found it harder and harder to listen to Robb's evidence that day. His impatience with Hardie's evidence had already drawn a rebuke from Turner Freeman's solicitor Tanya Segelov, who sat immediately in front of him.

'There was a running commentary from Bernie, like, 'How can you just sit there?' They'd say something and Bernie would say, 'That's rubbish!' There were times when I had to turn around and tell Bernie to be quiet because I couldn't hear through his commentary.'[61]

Months later Banton encountered Robb while he waited in a Sydney coffee shop for a negotiation session with Hardie. A young woman whose father was dying from mesothelioma approached Banton to congratulate him for his efforts in bringing Hardie to account, reducing him to tears.

'It was like a knife in the heart,' he told me. 'Robb was sitting at a table across the room reading a paper and he looked up to see me wiping a tear from my eye. I went over to him and thanked him, for at least partly telling the truth. He thanked me very much.'[62]

* * *

Four weeks after Robb's dramatic disclosure, Hardie gave the first public sign that it was shifting position. The company had commissioned a new actuarial study from KPMG which estimated the foundation had a half-billion-dollar shortfall.[63] Hardie blamed the 'flawed assumptions' of Trowbridge for the mistake. Now, it told the commission, it might be prepared to meet the shortfall provided the NSW government established a new statutory compensation scheme.

The following day the company left little doubt what would happen if its offer were not taken up: claimants would die with no compensation. A Hardie spokeswoman told a journalist: 'The alternative is potentially protracted legal actions ... They could take years and in the meantime the foundation would probably go into liquidation.'[64]

Already foundation directors were unable to obtain personal insurance for the potential damages they could face if they continued payouts knowing the funds could run out within three years. They sought an indemnity from the Supreme Court and during the next three weeks their case was referred to the Court of Appeal, with the real prospect that if they failed, the foundation would be declared insolvent.[65]

Combet was extremely concerned. 'Had the foundation gone belly up then people would have started dying in fairly quick order without any compensation, so I was desperately worried about it.'[66]

Around the country the clamour against the company grew louder. Local councils and unions began boycotting Hardie products. Banton and others appeared regularly on high-rating radio talkback shows with conservative commentators like Alan Jones attacking Hardie. International union affiliates mobilised to run a global campaign against the company, as the ACTU's Combet addressed union rallies and public meetings, labelling Hardie's latest offer of a statutory scheme as 'an attempt to blackmail the dying'.[67]

Combet described Hardie's actions to the conservative Sydney Institute (of which Hellicar was chair) as one of the 'most repugnant acts' in Australian corporate history.[68]

'They were trying to sell a stinking dead rotten cat. All the spin money in the world wouldn't help them,' he told me later. 'And they made a terrible mistake, too, in going into a bunker. It was just obscene. You couldn't possibly sell their position in a contested public argument. It was a disgrace.'[69]

At the eleventh hour, thirty minutes before the Jackson Commission ceased taking evidence, Hardie's lawyer Tony Meagher, SC, announced that the company would fund compensation for *all* those to fall ill from the group's asbestos products. He still failed to provide details about exactly what Hardie was proposing.[70]

One week earlier Hardie's board had appointed a new chair, Meredith Hellicar, who accepted the job after Alan McGregor was diagnosed with terminal cancer. She knew it would not be easy.

'I had two choices: I either walked away from the job and forevermore felt that I had let everybody down at a time when you needed somebody strong to do something. Or I stayed and did it, and had the broad public think I'm a shit.'[71]

Hellicar hired a new public relations company, The Third Person, and launched a blitz of media appearances in which she repeatedly apologised for underfunding the foundation and for the distress it had caused to asbestos disease sufferers: 'We are truly, truly sorry ... We are very, very sorry ... you cannot be blind to this awful, awful disease.'[72]

But she defended the actions of the Hardie managers and board. Hardie directors had an obligation to their shareholders, she said. Their legal advice had been that they were not permitted to contribute more funds to the foundation. And she insisted the NSW compensation scheme must be reformed.

Behind the scenes, Hellicar was once more trying to talk to Carr's government. 'I again personally made several approaches to the government to discuss that with them,' she told me. 'I said "I would prefer to put a submission in proposing a solution that would get support. Can't we at least discuss it with you?"' Her entreaties fell upon deaf ears. She was told that no meetings were possible.[73]

Banton's initial reaction to Hardie's eleventh-hour offer had been elation, telling journalists he was so overjoyed he nearly swallowed his oxygen hose: 'It took my breath away — and I only have 40 per cent lung capacity as it is!' His euphoria soon evaporated as he realised he had yet to see the fine print. He was even less impressed by Hellicar's public posturing. 'It makes me livid,' he told reporters. 'It's like moving the deckchairs on the *Titanic*.'[74]

Hardie had still made no guarantees and was pressuring Carr for a statutory scheme for asbestos compensation, something the premier had refused to rule out.[75] To counter Hardie's push the ACTU asked other state premiers to oppose statutory arrangements. Victoria's Steve Bracks and South Australia's Mike Rann quickly obliged. With Jackson's findings expected within weeks, national demonstrations were organised for the day of Hardie's annual meeting of shareholders, scheduled for 15 September 2004 amid heightened security at Sydney's Darling Harbour Convention Centre.

There, Bernie Banton addressed a 5000-strong crowd of supporters who had just marched through Sydney's streets waving placards and banners denouncing Hardie. Tens of thousands of others had joined similar rallies in cities throughout the country.

'James Hardie, I don't care how far you run or where you try to hide. Until they put me in a box I'm gonna chase you, and I won't let you out of my sight!'

The demonstrators roared their approval, and began to chant: 'Bernie! Bernie! Bernie!'

It was a feeling Banton never forgot. He took it personally. 'That was just so emotional,' he told me afterwards. 'All those people, chanting my name.'[76]

Behind him, the plate-glass windows of the Darling Harbour Convention Centre trembled slightly with the vibration from the noise. Inside the security cordon a small group of James Hardie shareholders gazed down at the scene before the information meeting. This was their first meeting since the Special Commission had begun. Meredith Hellicar began proceedings by expressing her 'profound regret' for the circumstances in which Hardie now found itself. But she also had a message of defiance.

'We do not accept,' Hellicar told her shareholders, 'that an outcome of this process should be the enrichment of plaintiff lawyers.' Hardie wanted changes to the asbestos compensation regime that would see legal and other expenses 'removed or considerably reduced'.[77]

At the demonstration outside I caught up with Banton just after his speech. He was rushing off for a TV interview, then on to Carr's office. The previous day he'd received a call: the premier was very keen to meet with him and some of his fellow sufferers that afternoon. At the rally he asked Turner Freeman's Armando Gardiman if he could organise for others to accompany him.

'I said, "When do you want them for?"' recalled Gardiman. 'And he said, "This afternoon at four o'clock." And I said, 'Fuck, you've got to be kidding!' That's when I hit the phone.'[78]

Within a few hours Gardiman managed to contact four other families with asbestos diseases to arrange their attendance. He chose them to represent a cross section of ages and asbestos exposure, hoping that the variety of their circumstances would drive home the message: that any compensation scheme required judicial discretion. There was no simple rule to fit all cases.

For Carr, the meeting was a turning point:

> That drove home to me the range and the diversity of the
> ways that people ended up being exposed … Someone spoke
> about picking up some slurry and bringing it back and using
> it as pouring for their garage. Then there was the widow
> who spoke about her husband erecting chicken coops in
> western Sydney with this material. They're the human stories
> that hook you. So it was one of those seminal meetings for a
> political leader, I guess.[79]

Eileen Day's husband, as a young man, had cut Hardie's Super Six asbestos cement sheets to build some chicken sheds; he had only died recently. According to Banton, Carr appeared quite affected by the family's story.

'Eileen's daughter really laid it on the line. She said how she missed her dad; what a terrible finish it had been, with so much pain. He was a big bloke, very active. She started to break down and that got everybody going.'[80]

The meeting lasted nearly an hour and a half, with the Labor Party caucus kept waiting in an adjoining room.

Over the next few days Carr decided upon an extraordinary course of action that would eventually result in a unique multi-billion-dollar deal for asbestos victims, the like of which had never been seen before. It was a strategy full of political risk, with the outcome never assured until the ink dried on the agreement signed with Hardie three years later. But for Hardie, the premier's newly found conviction was to prove decisive. Carr was to become a formidable enemy in a year that had already seen a spectacular unravelling of the company's fortunes.

There had been one jarring note at the meeting with asbestos disease sufferers and their families. As Carr was about to depart some of the waiting staff and ministers, including Attorney-General Debus and Industrial Relations Minister Della Bosca, joined the group. Debus later recounted that he then witnessed a 'noticeable level of aggression' from Cabinet Secretary Roger Wilkins towards the visitors.

'He was really quite belligerent. I remember it because I was surprised. We were all discussing, with great sympathy, possible future action and trying to find a way to move forward. He, to the contrary, had an aggressive tone of voice.'[81]

The ACTU Secretary Greg Combet, who had been talking with the ministers, also noticed.

'If anyone had been lobbied hard about statutory schemes, it was Wilkins. I can't remember exactly what he said, but he really got everyone's radar going. He didn't sound as if he was on board our team.'[82]

Unknown to most in the room, Wilkins had recently been spotted walking along Macquarie Street outside parliament engaged in conversation with Meredith Hellicar.

Hellicar categorically denied that the meeting ever took place. 'God, you fell for it, you read it in the media and you assumed it was true,' she said to me later. 'I remember, it was amazing that we were 'seen' together, it said. We weren't even together — I had not seen him all year.' She was adamant: there was absolutely no truth to the story.

'So you never met with him then?' I asked, just to be sure.

'I had not set eyes on him throughout all that year. There was no meeting at all with Roger Wilkins, at all, at any time during that year,' she replied emphatically.[83]

Wilkins, on the other hand, confirmed the encounter. He said it was a chance meeting.

I did run into Meredith coming out of Parliament House and walked with her for about a block on my way back to the office. She is a friend, not a close friend, but someone I have known for many years. For the record, there was no meeting about James Hardie or asbestos. There was not even a conversation about it, for that matter.[84]

The unexpected sighting set tongues wagging. Here was the state's most powerful bureaucrat talking, on obviously friendly terms, with the chair of James Hardie — a company embroiled in one of the country's biggest corporate scandals, and a company the government's ministers had publicly shunned.

Frantic inquiries were made by Banton's friends and a connection with the asbestos industry was discovered. Wilkins' father Bruce had worked all his life as an industrial chemist, first for Wunderlich, then Hardie.[85] The company's Camellia plant had been a familiar place for Wilkins, where he and his brother had done their homework after school as they waited for their father to finish work. Wilkins worked there himself, loading trucks with Hardie's asbestos cement planks and pipes during his university holidays.[86]

In the week following Carr's meeting, details of Wilkins' family connection and his meeting with Hellicar were leaked to Marcus Priest, a journalist with the *Australian Financial Review*. The day before the release of the Jackson Commission's findings his story was published.[87] Wilkins, declaring a conflict of interest, removed himself from any further government involvement with Hardie. If a conflict did exist, he said later, it was that both he and his father were potential claimants against the Hardie fund.

'I decided that there was no point in trying to figure out whether there was really a conflict of interest or not. There clearly was going to be grief for the government, for me and my family, if I did not stand aside. So I did.'[88]

He informed Carr. The premier told me later he did not treat it seriously.

'I don't think you can be judged by what your father does. He is a professional public servant. Would anyone seriously suggest that he bent the decision because his old dad, obviously in retirement now, had once before worked for James Hardie?'[89]

On the eve of the commissioner's report both the NSW manufacturers' union and its lawyers at Turner Freeman were again deeply suspicious of the Carr government's intentions. They had already fought against one attempt to wind back the asbestos laws. Hardie had made its offer conditional on such changes; Jackson had all but said acceptance of Hardie's suggestion might be the only practical solution to avoid the foundation's bankruptcy.

And the most powerful bureaucrat in Carr's government had been talking with Hellicar at a time his ministers had declared the company off limits.

But Premier Carr was about to have what he called his 'light bulb' moment.

twelve

STRIKING A DEAL

When David Jackson reported the findings of his Special Commission into Hardie's asbestos foundation on 21 September 2004, there was little ambiguity in the media coverage of the event. A CULTURE OF DENIAL screamed the headlines. JAMES HARDIE LIED. James Hardie and its executives had broken the law, engaging in misleading and deceptive conduct. Hardie's chairman, Alan McGregor, its CEO Peter Macdonald and general counsel Peter Shafron were 'difficult to believe'. Others, like the Trowbridge actuaries, had misled and their behaviour 'fell below the standards of professional care'.[1]

At first glance, it looked like the victory asbestos victims were hoping for. But while Hardie lost the battle, it might yet win the war.

Bernie Banton had been invited to join the premier, Bob Carr, to pick up the report at Government House. As he sat with his lawyers skimming the pages before the premier's media conference, a growing sense of foreboding gripped them all. 'The lawyers were incredulous,' remembered Banton. 'You could hear these comments as they read: 'Oh no!' and 'Gutless!'[2]

It was as they had feared. Although some of Jackson's conclusions were damning, others appeared bewildering. When James Hardie applied to move offshore, for example, Jackson found that both the company and its law firm Allens had breached their duty of disclosure before the Supreme Court, but the breach was 'not deliberate'.[3]

Greg Baxter, the company's PR chief, escaped any finding. Jackson had simply not considered his position due to lack of time.[4] Nor was Jackson prepared to find a breach of duty when Hellicar and the other directors cancelled the partly paid shares in the former Australian parent company — '... notwithstanding a lingering lack of enthusiasm for the commercial morality of the transaction'.[5]

Most worrying for Banton and his colleagues, Jackson identified a legal minefield over the critical question of how to force the Netherlands-based Hardie NV to pay up. It was as the company had predicted. Years of litigation would be needed to unravel this corporate restructuring. Meanwhile, the foundation would collapse and asbestos victims go without compensation. Jackson observed that the basis for a solution might well be found in Hardie's proposal for a statutory scheme, however much the unions and asbestos support groups might object.[6]

Premier Carr, though, had not been idle following his meeting with Banton the previous week. He had been concerned that, whatever the findings, his government would be blamed if it was forced to do a deal with Hardie. The secretary of the Australian Council of Trade Unions, Greg Combet, was keen to see the company brought to justice and had been staying in close contact. An idea took shape in Carr's mind:

I had an electric light bulb above my head one day that said:

'You've got the unions and the victims groups on the one hand, and you've got James Hardie on the other ... Why would we want to get involved in negotiating a settlement and appearing to own and sell a compromise?' We didn't want a compromise. We simply wanted to get a satisfactory package based on justice for the victims.[7]

Carr's solution was to hand negotiations with Hardie over funding and compensation to Combet, with a pledge that he would not sign off on a new compensation scheme unless the unions and victims groups agreed. The tactic was intended to maximise pressure on Hardie.

Combet agreed to Carr's offer with some trepidation, but he moved quickly, arranging a private meeting with Hellicar two days later. He was extremely anxious about the possible collapse of the foundation and he told her so: 'We've got to get the fucking thing sorted out, otherwise we're going to chase you to the ends of the earth. People will start dying soon without compensation. And it is untenable for the Australian directors and executives, in particular, to think that life is going to go on the same, because it's not.'[8]

Hellicar told me later that she acknowledged Combet would not accept any statutory capped scheme, but she suggested to him that a new compensation regime could reduce legal costs without capping payouts.[9]

The ACTU secretary chose his negotiating team. Although he had approached others from the asbestos groups, he decided that Banton was the stand-out candidate. But he excluded Turner

Freeman and the NSW metal workers' union from the negotiations.

'I didn't feel comfortable going forward with people who were wanting to attack Carr rather than focus on the enemy, which was James Hardie. So there was a fair bit of tension about that on our side of the campaign.'[10]

The union and its senior lawyer Gardiman saw it differently. They had no grudge against Carr: they had just seen Hardie steamroll too many people before and were concerned that in the end it would be asbestos victims who would lose. For Banton it was a tricky balance. His personal lawyers were Turner Freeman, yet it was now only Slater & Gordon's Ken Fowlie who would continue to run the legal side of the negotiations. Combet's suspicion was reciprocated by Gardiman, who wrote to his asbestos clients warning them that key tribunal awards, like provisional damages, might be traded away in the negotiations.[11]

When the talks began Hellicar appeared for the cameras, but after they had left said that Dutch law prevented her further participation. Banton recalled how she opened the meeting: 'The first thing, Bernie, I want to say is how very sorry I am for what has happened ...'

Banton was unmoved: 'I take all that with a grain of salt,' he replied. 'Words don't mean much. Let's see how your actions are in eighteen months.'[12]

Along with a lawyer from Allens, Hardie had chosen Peter Hunt, senior partner of the financial advisers Caliburn, for its team. Hunt began explaining Hardie's achievements and Banton took an immediate dislike to him.

'I put up with it for about a minute, then I just exploded. 'They've killed thousands of people — and they knew!' I went for this bloke — and from then on every time he opened his mouth I just went for the jugular.'[13]

Undeterred, Hunt took to phoning Combet between meetings, on one occasion calling seven times on the same day. He failed to impress the ACTU secretary. Although Combet had drafted a broad one-page in-principle agreement for how negotiations should proceed, within a few weeks it was evident they were going nowhere fast.

'I just couldn't cope with dealing with Allens. They were just so up to their necks in the restructuring that I wasn't prepared to deal with them and I told Hardie that. This guy Peter Hunt from Caliburn even tried to make friends with me and position himself as the lead guy, and I didn't take much of a fancy to him either.'[14]

Leon Zwier, from the Melbourne law firm Arnold Bloch Leibler, was one person representing Hardie whom Banton and Combet learnt to respect. Combet had dealt with Zwier before and trusted him, and had recommended him to Hellicar at their first meeting. Banton said later that Zwier made a big difference over the tricky financial negotiations that followed.

Another Hardie negotiator was John Atanaskovic, a corporate lawyer with a reputation for hard dealing. Combet was particularly frustrated that he had no person from Hardie itself to deal with directly.

'The whole thing gave me the shits. You always have to negotiate with principals. You can't negotiate with advisers, they don't have the authority to make decisions.'[15]

But Hellicar had problems of her own. One week after Jackson had handed down his findings, the two men he had singled out for misleading and deceptive behaviour, the CEO Macdonald and former general counsel, now chief financial officer, Shafron, were still in their jobs. When they eventually stood aside they remained on the company payroll, with Macdonald continuing to run the US business where most of Hardie's profits were based. It took a month for the two men to resign and even then Hardie retained Macdonald as a consultant.

Flanked by her advisers, Hellicar walked confidently into the glare of TV lights at the Meriton Hotel near Sydney's Circular Quay to field questions from the media about their departure.

'Mr Macdonald and Mr Shafron have resigned ... I believe they both intend to vigorously defend any allegations that might be brought against them ...'[16]

Would the company be paying for their defence? asked a journalist. Hellicar conceded that under Hardie's indemnities and insurance policies some defence costs would be paid.

Why did Macdonald remain as a consultant?

'Mr Macdonald ... is responsible for the incredibly successful performance of the company,' she said. He would advise Louis Gries, his interim replacement. Russell Chenu had taken Shafron's place. The disgraced executives were to be paid handsomely. For Macdonald, it was US$6.5 million plus an ongoing monthly retainer of $77,000. Shafron received US$865,000.

Not surprisingly, the huge payouts for both men dominated the headlines the next day, particularly given that the corporate regulator, the Australian Securities and Investment Commission (ASIC), was now investigating their behaviour, with the possibility that charges could be laid.

Among the barrage of questions directed at Hellicar I managed to ask one about Reid.

PEACOCK: Have you spoken recently to your major private shareholder, John Reid, the former chairman of the company?

HELLICAR: Not very recently. But certainly as a friend and colleague we have spoken this year ... I'm not going to talk about a private conversation with somebody who I was speaking to as a personal friend.

Behind the scenes, it seemed, her mentor and adviser was still active. John Reid, the man who refused to speak publicly about the asbestos victims from his years of running the company, was still in touch.

Hellicar's summary of Jackson's findings made it seem that Hardie had done nothing wrong:

'... the intentions of the board that I was on at the time was to properly fund asbestos victims; that the establishment of the foundation was legitimate; that the move to the Netherlands was legitimate; that this company has no legal liability to make payments ...'

She evidently was convinced that Hardie was still negotiating with Combet from a position of strength. It was.

Hardie's media conference lasted about forty minutes. Hellicar had carefully avoided answering questions about another crisis unfolding behind the scenes. Two days before, the foundation had warned her that its financial situation was so dire it might soon be forced to apply for the appointment of a provisional liquidator. It released the news publicly a few hours after her media conference.

The foundation had just over $40 million in the bank and although Hardie had offered to advance more, its offer was still conditional on agreements to indemnify the company against legal action. For the foundation's lawyer, Nancy Milne, these conditions were completely unacceptable. Hardie's lawyer, John Atanaskovic, accused her of misreading the offer. Milne stuck to her guns. She said later there was 'a lot of pressure', with Atanaskovic 'trying to split hairs' over the wording of the conditions.

Not all the foundation directors wanted to launch proceedings to call in a liquidator, least of all its chairman, Sir Llew Edwards.

'It's a pity this story is being beaten up in this regard,' he told me at the time. 'This is only one option that may occur if we're unsuccessful in bringing people to the table to provide additional funds.'[17]

But his legal advice was clear. The additional funds on offer could not be accepted if they were tied to a legal indemnity for Hardie NV or its erstwhile Australian parent JHIL. The foundation duly filed its application for liquidation and the matter was listed for the following month.

By now Combet and Banton had begun to meet regularly with Hardie negotiators at the offices of Lazard's, the investment bank, high above Circular Quay overlooking Sydney Harbour. Both were increasingly frustrated with the snail's pace of their talks. Combet became convinced Hardie was stalling. But the public campaign and Banton's frequent media appearances were taking their toll on the company's share price.

A Hardie insider told me later: 'Bernie really upset them. They'd watch him, and speculate if he was genuinely ill or not.

'That fucking Bernie! What a liar! It was CSR as well,' they'd say [a reference to CSR's joint ownership of the asbestos factory where Banton had worked]. And I'd be sitting there, thinking the guy's a hero, he's got two hoses up his nostrils ...'[18]

Combet and Banton were generally seething following their negotiating sessions with Hardie. 'Grubs!' 'Worms!' read my notes of Banton's comments. 'I feel like a shower after meeting them,' Combet told me. But as public pressure continued to mount he sensed the tide might be turning. Banton's media campaigning was having its effect, as was Carr's uncompromising position. According to Combet, progress was dogged by Hardie's insistence on a statutory compensation scheme:

> It seemed to be off the table but it kept coming back onto the table. It was the slipperiest, sliding, slimy type of negotiations ... It just gave me the fucking shits! I had to keep suspending the negotiations because every time they came back with the response to our document, they were trying to reintroduce a statutory scheme by some sneaky manoeuvre. They must have taken me to be a complete fool.[19]

The ACTU secretary was at least grateful that someone from the company was now directly participating in the talks. Russell Chenu, the newly appointed CFO, had embarked on a steep learning curve to acquaint himself with the history of the foundation. Combet found he was 'decent and straight', although he still required time to get on top of the subject. The advantage was that, unlike the others, he took Combet at his word. When the ACTU secretary said he would not accept a statutory scheme, he meant it.

'I knew what the game was here — an attempt by the company to severely restrict people's compensation rights and their compensation payment. And I was not going to be part of it.'[20]

Apart from any personal conviction, Combet had another reason to be wary. He had already fought with Turner Freeman and the NSW manufacturers' union about his trust in Carr on this very issue. Any sign that he was about to cave in to Hardie's pressure would prompt them to go public with their concerns. With a national election looming, in which a key battleground would be workplace laws likely to wipe out unions, Combet needed a unifying victory, not a damaging division.

Premier Carr, in any case, was also applying his own pressure on the company. He sent the US Securities and Exchange Commission a copy of the Jackson findings, asking it to investigate Hardie's behaviour.[21] The premier also announced that his government was considering special legislation to unravel the Hardie's move to the Netherlands.

The public campaigning and private negotiations were taking their personal toll. Combet's daughter refused to talk to him for months after he left a school concert early. And Banton's asbestos condition was worsening. He became increasingly breathless. His lung function disability was soon reclassified at the Dust Diseases Board from 60 per cent up to 80 per cent impairment. His wife Karen, who dropped him off for the meetings, some of which used to last several hours, said she noticed how they affected him.[22]

A way around the compensation impasse suddenly struck Combet during a frustrating weekend in Melbourne when he had been fretting about what to do. He would remove it from the table. Combet was inspired by a TV story on negotiations in the Middle East, when Israel announced a unilateral withdrawal from Gaza and caught Palestinian negotiators by surprise. 'It just sort of came to me that we needed a "Gaza solution",' he told me, 'where I just withdrew from the compo thing altogether.' The next day he explained his plan to the premier's staff. Could the government review the costs and recommend improvements to the tribunal's procedures while still preserving the common law rights of the asbestos victims? Carr agreed.

The ACTU secretary chuckled when he recounted his next meeting with the Hardie negotiating team. It was one of his most enjoyable moments during the campaign. The lawyers had begun to discuss the labyrinth of compensation details when Combet cut them short with the announcement that he had arranged for the government to review the system. He introduced the deputy director-general of the Cabinet Office, Leigh Sanderson, who was waiting outside the room, and she explained the government's decision.

Combet recalled the moment with evident relish: 'Hardie looked absolutely like shot ducks. I was just able to say, "Well, that's that. You make a submission to the review and away we go on with the commercials. Let's focus on the main game."'

The next day Carr announced a 'short, sharp review' of asbestos compensation arrangements, which would focus on reducing costs while preserving victims' rights. It would not, he emphasised, be a statutory scheme. Hardie had little choice but to grin and bear it and Hellicar issued a statement welcoming the review.[23]

It proved to be a pivotal moment in the negotiations. With the compensation issue off the table, Combet and his team at last began to make headway towards an agreement. It could not come a moment too soon: another crisis was looming at the foundation.

After a month's bitter stand-off, during which Hardie's former Australian parent continued to insist that a legal indemnity would apply if the foundation accepted its accelerated payments, the foundation was due to refile documents for possible liquidation. Payments to claimants would then cease and it was possible that future claimants would receive no compensation whatsoever. Hardie transferred a first installment of $31 million overnight into the foundation's account. The foundation sent it back. It was not prepared to forgo future legal action against Hardie companies.[24]

Premier Carr applied more pressure. Hardie was 'wrecking its image for all time' if it didn't get the message and bring the negotiations to a conclusion, he warned.[25]

Hellicar returned fire against the barrage of bad publicity. The company had engaged Jackson Wells Morris (JWM), another expensive media consultancy with impeccable conservative credentials. The firm's founder, John Wells, had been an adviser to previous Liberal Party leader Andrew Peacock, and recently had been joined by Grahame Morris, a longtime adviser to the current prime minister, John Howard. JWM designed a media strategy for Hardie with the key communications messages it would need to 'climb out of the bunker'.[26]

The corporate image was again revamped. Hardie would present itself as the injured party, misunderstood and maligned by others with vested interests, despite its best efforts to do the right thing. Hellicar was soon explaining to ABC Radio that the company had suffered because it had failed to present its case properly.

'We can't help it if we're not legally liable — we did not rearrange our affairs to avoid legal liability and the commission confirmed this ... Our integrity is intact. There's been nothing wrong with our actions. We have made some bad errors in communications.'[27]

The problem, she asserted, had been one of presentation. She was astounded by the mythology that had developed.

'James Hardie has not been in the business of spin. We've been in the business of being battered ... this company is run by real people full of the greatest integrity.'

Hellicar staunchly defended the former CEO Macdonald, whom Commissioner Jackson had found to lack credibility under oath: 'This man is not evil incarnate. Our shareholders are very concerned about the loss of his knowledge.'

The morning the foundation went to court for liquidation there was a dramatic change in the attitude of the influential conservative radio talkback host Alan Jones. Until then Jones had been a vigorous supporter of Bernie Banton, just as in previous years he had been highly critical of Carr's amendments to liability laws. But this day he presented a different message for his listeners as he interviewed Hardie's chair.

Why wouldn't Hardie guarantee there was always enough money in the jar for claimants? Jones asked Hellicar.

'We did that last week,' she answered.

So why was the foundation applying for liquidation?

'Alan, honestly, I just don't understand,' replied an apparently distressed Hellicar. 'I am just sick and tired of the stunts and politics of all this. I think you made a comment this morning,

'Why is this being played out in the media?' ... I genuinely don't understand.'[28]

The Hardie chair continued her media counterattack, projecting an image of reasonableness.

'We were trying to talk to Greg Combet all day yesterday and I was amazed to find television footage last night that he was in Sydney holding press conferences but not returning our calls to sit at the table. It's ridiculous. We want to get this done ... The premier has refused all year to meet with us. He refuses our letters ...'

Jones concluded with a short commentary: either Hellicar was lying or she'd been totally misreported, and he didn't think she was telling lies. 'There are some people, it seems, in all of this who are kicking the football around ...'

I was driving to work at the time Hellicar was being interviewed by Jones and was blissfully unaware of the broadcast. I normally didn't listen to his program. But I was soon to hear about it.

My mobile phone rang soon after. It was Carr's office. The premier came on the line. It was the first time he had called me since taking office.

'Did you hear Jones?' he asked urgently.

Still adjusting to the idea that he had called me, for a moment I had no idea what he was talking about. Jones who? Then I realised.

'It was amazing,' the premier said. 'He had Hellicar on. He's done a 180-degree shift.'

I considered for a minute and said, 'Yeah, well, they have a new PR company. Grahame Morris and John Wells.'

'Ah,' said Carr thoughtfully. 'That probably explains it.'

We spoke a few more words, then he rang off, leaving me to ponder the unusual call. I was intrigued by the fact that Carr was so sensitive about the talkback show. And why call me? It suggested that Carr was extremely nervous about how the story would continue to run.

Years later, Carr remembered the moment: 'There was an overnight change. He [Jones] swooped in and took up the case of defending this company root and branch. This was someone who'd taken the cause of the plaintiff lawyers in the tort law reform process. Someone got him on to that. He had her on and she was saying 'the Government won't even talk to us'.'

He laughed, looking back. 'That was very, very deliberate.'[29]

At the time, though, he was not laughing. It was a tense few weeks. Nobody knew whether the foundation might fold. If it did, as Combet feared, anything could happen. The foundation's solicitor, Nancy Milne, was furious. Not only did she feel hectored by Hardie's lawyer Atanaskovic, now she had heard Hellicar on the radio, in her view totally misrepresenting the situation.

'These stupid people, what are they doing? Here we are offering them the money and they won't take it, scaring all these victims ...' she mimicked.[30]

Hardie's media campaign escalated. The company took out large newspaper advertisements headlined: THERE ARE TWO SIDES TO EVERY STORY. A fortnight later Hellicar was back with Alan Jones, who told his listeners he had started to 'smell a rat'.[31] But he warned the Hardie chair it would not be a 'soft' interview. The negotiations were still dragging on. Was there a draft heads of agreement shared by the parties on 3 November? he asked

Hellicar sternly. She confirmed there was. In fact, it was Hardie's own draft that it had sent to the ACTU.

Jones took her through some of the details. Since that date, Hellicar told him, the ACTU had failed to provide a detailed response. 'And look, we acknowledge this is complex, but the problem is ...' One could imagine her shrugging with exasperation.

> JONES: But your submissions are consistent with the agreed terms in the heads of agreement?
> HELLICAR: Well, the heads of agreement itself hasn't been agreed.
> JONES: But it's consistent with what you think ...
> HELLICAR: Yes.
> JONES: ... the heads of agreement should be agreed to?
> HELLICAR: Absolutely.

Jones was asking her if Hardie's latest submission was consistent with the company's own proposed agreement. It wasn't a soft interview. It wasn't even an interview. It was a free kick.

The talkback host continued. Where was the NSW premier during all of this? he wanted to know. Had Mr Carr spoken to anyone at James Hardie?

> HELLICAR: No, he won't. He refuses to meet with us.
> JONES: How can he make judgements about what James Hardie are or are not doing if he refuses to speak to James Hardie?
> HELLICAR: I have no idea, Alan.

The interview concluded, as before, with a pithy summary from Jones: 'That's Meredith Hellicar. And that's a disgrace ... where does the New South Wales government stand in all of this? And I think victims need to understand that some people are playing egos here, but not reality!'

That was certainly not how the most famous victim saw it. A few hours later my phone rang. This time it was Banton, hopping mad. He had been in negotiations the day before and was due back there shortly.

Banton fumed: 'Whenever we take two steps forward, James Hardie takes two steps back. Did you hear Alan Jones this morning? 'The ACTU has not responded to the heads of agreement.' That's a straight lie! As late as last Friday, Saturday night, Greg and Leon Zwier updated the agreement.'[32]

Banton hit the news bulletins again that evening, commenting that '... after the spin on Alan Jones this morning I just wanted to vomit'.

But behind the public posturing the two parties were closer than they had ever been. Chenu and Combet were inching towards an agreement. The foundation's liquidation proceedings were adjourned until January to allow time for a resolution.

Carr tightened the screws again. Earlier he had legislated to grant ASIC full access to the documents provided to the Jackson Commission, overriding Hardie's professional legal privilege. Now he granted the same access to another federal agency, the Australian Competition and Consumer Commission.[33]

Hardie directors provided a last-minute impediment to the negotiations, insisting the union bans and council boycotts against Hardie products be lifted and that all civil claims against themselves would be extinguished.

Three days before Christmas the heads of agreement was signed, the largest personal injury settlement in Australia, that was to last at least forty years. It was a first step, not legally binding, leaving the principal agreement with its fine detail to be sorted out.

Banton tersely summarised the mood: 'Hardie? I don't really have one word for them. I'd just like them to hurry up and pay up.'[34]

For now, at least, asbestos victims from the Hardie group had reason to hope. But their nightmare was still far from over. Within six months nerves were once again fraying.

The foundation lived a precarious existence during these months, as its directors continued seeking court protection to go on making compensation payments until a final deal was in place.

The NSW government's review of the compensation system was completed by July 2006 and new procedures for claims before the Dust Diseases Tribunal came into effect.[35]

Hardie portrayed this as a vindication for its contention that tribunal procedures had been eating up the lion's share of victims' funds. Most of the delays and costs in the tribunal, in fact, were the result of cross-claims in which Hardie and other defendants argued over what proportion each company should pay of the compensation due to an asbestos disease sufferer.

The government's new claims resolution process sought to encourage both claimants and defendants to provide each other with details of each case as early as possible. Mediators were to facilitate speedy agreements. Early signs were that the new process did indeed reduce costs, but within eighteen months a worrying picture would emerge which prompted criticism from judges, plaintiffs and defendant lawyers alike that the new scheme was little better than the one it had replaced.

Meanwhile, negotiations over the funding deal dragged on into the ninth draft, bogged down by the complexities of Hardie's accounting, tax and multi-jurisdictional nature. NSW Premier Bob Carr resigned. His successor, Morris Iemma, signalled his displeasure with the delay.[36]

If the new premier was impatient, Banton and the asbestos groups were even more so. Their fear — that without a legally binding agreement the company could still walk away from the deal — began to grow again. Unions threatened to recommence their national campaign if the company reneged.[37]

Another demonstration was held outside the James Hardie shareholders' information meeting. Inside Hellicar admitted the agreement had taken an 'appalling' length of time but, she cautioned, 'we need to ensure we take all the care and time in the world to get this as right as possible'.[38] Hardie wanted tax deductibility for its payments to the foundation; its directors had also demanded indemnity from civil suits, something Carr had been prepared to promise but Iemma wasn't.

Inside the government, a senior bureaucrat observed that Hardie's behaviour remained the same as it had always been. 'They just didn't give an inch on anything,' he said. 'Everything was all about absolutely minimising the amount of money they could find or afford.'[39]

Three months later the two parties moved into their eleventh draft, with the premier and Hardie trading public shots about the delay.

Combet warned that time had run out for the company: 'It's eleven months since we concluded the heads of agreement and that's long enough for the legal documentation to be prepared. But James Hardie is still trying smart-alec lawyers' tricks.'[40]

The NSW government drafted legislation to wind back Hardie's move to the Netherlands and force it to pay. According to the NSW minister John Della Bosca:

> James Hardie played it tough right to the end. Right up until the last minute. We were pretty worried that it was going to go off the rails. We were literally backing and filling it until two days before we put the legislation through. We had a choice and they weren't going to put us off. We were either going to do the nice legislation or the ugly legislation.[41]

Two days before the legislation was due to be introduced into parliament, Hardie and the government announced that substantial agreement had been reached, and all that was required was approval by the company's board. That came at the end of the week.

At the joint media conference in the premier's offices that Friday, journalists milled impatiently. They were kept waiting an hour.

When the parties finally signed the document, Hellicar beamed and planted a kiss on Banton's cheek as the cameras flashed.

Combet stepped back, aghast.

'I think that was her media advice and I just wasn't going to be party to it. I didn't feel like kissing her. I didn't want the image of her kissing me to be anywhere, so I reeled right back. I wasn't going to be in that!'[42]

The following day the *Australian Financial Review* reported a reason for the delay to the signing ceremony. According to the journalist Marcus Priest, 'right down to the wire, the company fought to change the deal', and at the scheduled time to sign the agreement, 'Hellicar was still refusing ... until a clause was changed to give the company more time to sue unions and asbestos victims over past protest actions and boycotts'.[43]

Hardie immediately demanded an apology, describing Priest's story as absolutely untrue and defamatory, and asserting that the delay had been caused by the government's insistence on a last-minute amendment. The newspaper published a correction in its next edition. But it was Banton who had given Priest his information. Incensed, he wrote to the newspaper asserting the original story was true.[44]

As another year rolled by it became apparent the asbestos deal still wasn't locked in. It was 2006, and several conditions still had to be met before Hardie would approve the arrangement, the most important of which was securing tax-deductible status for money paid into the foundation's successor, to be called the Asbestos Injury Compensation Fund. The Australian Tax Office advised that Hardie payments to the fund would not be deductible.[45] Yet another round of negotiations began.

Midway through the year, the wrangling continued. This time the dynamics were slightly different: Banton, Combet and the NSW government had all joined James Hardie in urging Prime Minister Howard's federal government to change the law to allow payments to the fund to be tax deductible. Neither John Howard nor Treasurer Peter Costello was sympathetic. Banton's health had started to decline, and he told a reporter the negotiations had taken their toll.

'This has been so flaming tough. It's been a hard slog and I'm worn out, it's as simple as that. I am really running out of petrol. This has taken an enormous emotional toll and the longer it goes on, the harder it is.'[46]

Emotions began to run high in August when Hardie directors sought to double their fees. Hellicar's payment was to increase from $235,000 to $394,000 a year; her future retirement package was also to leap. Sydney's *Daily Telegraph* sent a reporter to the company's AGM in Amsterdam to hand out petitions urging shareholders to block the pay rise;[47] demonstrators outside the company's Australian meeting branded it immoral, while inside one shareholder called it 'obscene'.[48]

Life at the foundation had not become any easier. After hanging on by its financial fingernails to survive until the deal was struck, it now found itself waiting another year before funds would begin to flow. Again, alliances had been shifting. Just as Hardie had expressed its distrust for how the foundation's directors had managed its affairs, so too did the NSW government distrust the board Hardie had originally chosen to run it.

As discussions developed between Hardie and the NSW government on the shape of the special fund to replace the foundation, both agreed they did not want the existing directors to remain. On the eve of the heads of agreement in 2004, the attorney-general, Bob Debus, informed the foundation's chairman, Sir Llew Edwards, that he would like the directors to resign.[49] Privately, Debus had informed Ian Hutchinson that he wanted him to remain for the transitional period to the new fund.

But Edwards believed that if he was to go, they all should go. 'We're all in this together,' he told Hutchinson. 'You should resign with us to protect our reputation.'

'Sir Llew, nothing I do will protect your reputation,' answered Hutchinson.[50]

Their relationship, which had been brittle enough up until then, deteriorated dramatically.

Edwards dug in his heels. He knew the attorney-general had no immediate power to sack him. He asked for talks, but Debus saw no point. The stand-off continued until February the following year, by which time the government had passed legislation enabling it to wind up the foundation. Unusually, it also empowered the attorney-general to appoint directors to the existing company. Debus repeated his request, adding that if Edwards and his colleagues preferred, he would have them removed.[51] Separately, he again asked Hutchinson to remain.

By March, the last of the Hardie-appointed directors — Edwards, Jollie and Cooper — had resigned. The government appointed Nancy Milne, the foundation's solicitor, as a director to join Hutchinson, as well as a former partner from his legal firm, Freehills.

One difference to the foundation's affairs became immediately apparent. A saving was made of nearly half a million dollars in directors' fees.[52] But the foundation's finances continued to be in a parlous situation, confirming Commissioner Jackson's earlier prediction that it was likely to run out of money by 2007. It had fallen so short of cash by December that a Hardie NV company provided a loan to ensure its survival until the transition to the new Asbestos Injuries Compensation Fund, which began operating two months later.

There was an additional significance to the delay caused by the interminable negotiations. Although Commissioner Jackson had concluded that legal avenues for the foundation to obtain the missing funds from Hardie were fraught with possible delay and cost, even the remote possibility of a successful action would expire six years after the foundation's creation, due to the statute of limitations.

To be ready, the foundation arranged for summonses to be drawn up for issue against Macdonald, Shafron, Morley and other Hardie executives, with more to be added should the action be launched.

'15 February was "high noon",' Hutchinson recalled later. 'If the shareholders had not approved the revised agreement at the Amsterdam meeting on 7 February, we were ready.'

But the deal was approved only days before the deadline expired.

There was elation when eventually the new arrangements came into place. It was, Banton remarked, a great victory, with a deal unlike any other negotiated anywhere else in the world. Most of the negotiating team, though, simply felt exhaustion and relief after a long and tortuous struggle.

Combet was sure he'd done the best deal he could: 'I fought tooth and nail to get what we got, mate. Absolutely fought as hard as I possibly could and I don't believe I could have got more.'[53]

In the end Milne believed it was media attention which made the difference.

It was a media thing. There just became this huge groundswell of public opinion which really drove the outcome. The fact that you'd have front page of the *Australian*, the *Herald*, the *Daily Telegraph* — all talking about asbestos scandals, the lies — that was really the issue, I think, that forced the resolution. And that was what I think we had always felt in our bones. You might not have a legal solution, but the solution was to put it all out there and hope that the media did get hold of it and would push.[54]

And for the media, it was Bernie Banton who had made the difference, as an editorial in the *Sydney Morning Herald* headlined NO GLORY FOR JAMES HARDIE concluded: 'Over the many years of difficult negotiations ... Mr Banton has given quiet but compelling voice to the just demands of thousands of asbestos victims. An ordinary bloke, seriously ill and relying on an oxygen bottle, Mr Banton accepted the challenge of extraordinary times. Now there is a man of which all Australians, can, indeed, be proud.'[55]

But this was not the end of the saga.

For some of the sufferers of asbestos disease caused by Hardie's former companies, the NSW compensation regime may have changed for the worse. The 'plaintiff-friendly' tribunal, which Hardie had sought to marginalise before deciding to separate from its asbestos companies altogether, ceased to be the international model it had once provided. A little over a year after the new claims resolution scheme was put in place, both plaintiff and defence lawyers were complaining of 'micro-management gone mad'. The new procedures applied over sixty time limitations, governed by seven separate acts, rules or regulations: no other NSW civil court had more than four.[56]

According to James Sheller, an occasional barrister for Hardie's former companies, the new regime was 'bureaucratic, pedantic and duplicative' and almost certainly cost more.[57]

More worryingly, the scheme appeared to be failing those in greatest need, the 50 per cent of asbestos disease sufferers who contracted mesothelioma. According to Gardiman, in the six months to March 2008, nearly sixty plaintiffs died before their claim had been resolved. 'It is worse than it has ever been in my experience,' said the lawyer. 'The system is failing and needs to be fundamentally reformed.'[58]

Judges, too, complained from the tribunal bench and urged the government to make changes. Observed Judge Bill Kearns: 'There is something seriously wrong with a system that, because of its structure, denies a plaintiff access to a judge until ... the last few days of his or her life and at a time when he or she is in a state of seriously poor and deteriorating health and with little prospect of his or her case being completed within his or her lifetime.'[59]

An ironic aftermath of the bureaucratic intervention is that lawyers from both sides of the asbestos divide, plaintiffs and defendants, began to negotiate settlements directly, without the 'bureaucratic nightmare' they say the new procedures involve. In that sense, the outcome has been positive.

All the while, asbestos hazards in Australia's midst continued to linger, while those with the knowledge and responsibility for their existence continued to keep their secrets. Hardie had monitored some of these hazards for more than three decades. With the new deal, it had won a legislative immunity against claims for some of them that was denied to other asbestos producers. The company might yet have the last laugh.

thirteen

UNDER THE CARPET: UNFINISHED BUSINESS

In late 2005 the embattled chair of James Hardie, Meredith Hellicar, spoke warmly to me of a letter of support she'd received from an elderly woman. 'This wonderful ninety-three-year-old woman ... was married to two James Hardie plant managers in a row,' Hellicar said. 'She said they both loved asbestos. One of her husbands lined their driveway with asbestos.'[1]

For Hellicar, the letter provided reassurance that she was continuing an honourable company tradition set by her predecessor John Reid and his family, one that reflected the best of moral corporate behaviour.

But what neither Hellicar, nor anyone else from Hardie, ever said publicly was that such innocuous-looking driveways might kill. They are yet another part of the deadly legacy kept secret from an unsuspecting public by a company determined to minimise its legal liabilities. The corporate culture of deceit identified by the Jackson Commission developed many decades ago and persists to this day. Hardie's victims will continue to accumulate because the company has never told the full truth about the asbestos hazards left in its wake. Some, quite literally, have been swept under the carpet.

Until the 1970s it had been common practice to build such domestic driveways, paths and garage floors using Hardie's asbestos waste. The company encouraged its employees to help themselves to the 'fines', as it was called.

'People loved it! They put it in their driveway. I did it myself. To us it was cement,' a former executive told me. 'You'd just lay it out and hose it. There'd be hundreds of houses around with it, because it was very popular.'

Neil Gilbert, the former Hardie engineer who established Hardie's dust extraction system and medical surveillance scheme before quitting in 1971, admitted to worrying about the driveways, but felt that little could be done. 'There were so many of them. We did nothing about it because there were so many of them. I took it home myself. It was bloody good. It would be futile to try to chase up all the cases. It was absolutely everywhere. Employees would turn up with a trailer. I sold broken pipes for charity and built a scout hall with the proceeds.'[2]

The compacted waste still remains in people's driveways. I have seen one in a quiet suburban street bordering the old industrial area of Elizabeth in Adelaide, where the Hardie factory had produced its asbestos pipes and sheets. It had the appearance of concrete, with its grey colour and hard surface. Only a close examination at the edges revealed the telltale fraying fibres, glistening among the grime.

According to Gilbert, thousands of driveways would have been built this way, but he shrugged at the suggestion that something should be done about it. 'That's fate,' he told me. 'I know where there's one or two. It would take a massive amount of publicity to track them down. Most people wouldn't recognise them. You couldn't tell the difference between it and concrete.'

One of Hardie's biggest fears has been that litigation against it could extend beyond simply paying for the deaths and injuries it has caused, to cleaning up the dangerous materials it has left behind.

'The establishment of a broadly defined duty to remediate, whether at common law or by statute, could have a catastrophic effect on the company,' warned Hardie's litigation manager Wayne Attrill in 2001, just before the company shed its asbestos subsidiaries and moved offshore.[3] The prospect of such claims for remediation also featured in the early discussion held by Hardie's counsel Peter Shafron and Michael Gill when they first canvassed the idea of separating the company's asbestos subsidiary.[4]

In the protracted battle with the NSW government which followed, Hardie extracted unique legal protection from just such a duty.

No one knows how many people have been exposed to lethal doses of Hardie's asbestos waste. What is certain is that many thousands could have been; what is equally certain is that the company has known that their lives have been endangered.

For Hardie, the driveways were part of a bigger problem. Thousands of tonnes of its asbestos waste were dispersed in all sorts of places: in rivers and creeks, on vacant blocks, on roadways, even on football ovals. Wherever fill was needed, the Hardie's waste was available. The practice was made even more dangerous because a large percentage of the waste came from moulded products like pipes, which had contained the deadly brown and blue asbestos. As a Hardie memo in 1977 about its Camellia factory in Sydney noted: 'It is understood that our dust was a sought-after item and was even sold. It was particularly useful for light duty paths, garage floors and general filling. Our reject and broken scrap was also very useful as a filling for driveways, etc. in many of the market gardens west of our factory.'[5]

Father John Boyle, whose father worked at the factory, remembered the asbestos driveway and garage floor from his family house near Parramatta. His mother Molly had helped his father lay the asbestos waste and for years later used to sweep the garage floor clean:

> I know well what it looks like. It's a fibrous, powdery material
> that along with water becomes as hard as concrete. It was a
> cheap fill, and in those days it probably wasn't seen to be so
> bad. It's good for ten or twenty years, but then it breaks up,
> and that's when the fibres are released. From the mid-seventies
> James Hardie knew about them but didn't warn people.[6]

In fact Hardie had known long before that, but it was only during the 1970s that it began to do anything about it. The company's reaction, as Boyle noted, was not to alert people to the dangers; instead, it set about quietly stopping the practice.

John Boyle's mother died from mesothelioma. She had also been exposed to asbestos from her husband's overalls, but it seemed likely that her greatest exposure was from the driveway and garage floor. Hardie settled her compensation claim.[7] Remarkably, it also paid for the removal of the asbestos waste from the Boyle household, but as the lawyer Attrill noted in 2001, this was then the only remediation claim the company had met, and was one '... which arose out of a peculiar set of circumstances'.[8]

Those 'peculiar circumstances' were that Boyle's father had been the best mate of Alan Overton, Hardie's former transport manager who, although retired, was on a retainer to the company because of his extensive knowledge about the company's waste disposal. Overton had threatened to give evidence on behalf of Boyle. The company soon agreed to a settlement which included the cost of removing the asbestos.

Not so fortunate was the suburban solicitor David Gleadall, who had never set foot inside a Hardie factory. His father had helped a local contractor modify his truck with a special scoop designed to pick up and distribute the Hardie waste, and had arranged for a load to be dumped at the family home in Dundas in Sydney's west.[9] As a child, David helped him smooth the asbestos tailings across the new garage floor and driveway, and one of his chores was to sweep the floor clean. At age fifty-one he developed symptoms of mesothelioma, although it took two years before doctors correctly identified the cancer. The plight of Gleadall and his family made a deep impression on Premier Carr when he met asbestos victims in 2004.

Children were also to fall victim to Hardie's poisonous landfill. Anthony Cini grew up in Sydney's Pendle Hill, not far from the Camellia plant where his father worked. Part of a large Maltese family, he remembered helping his dad load up his truck with asbestos waste which they spread on a driveway around their house and on pathways through the market garden his parents established. 'My father, brothers and I used shovels and rakes to level out the asbestos tailings on the ground,' he told his lawyer. 'My brothers and I played with the asbestos ... we rolled up balls of asbestos dust and fibre and threw it at each other.'[10]

He contracted mesothelioma when he was forty-six. Before he died, he feared for his family. 'I'm concerned and worried that my mother, brothers, or sister might get what I have too. All my family to some extent or another worked, handled or inhaled asbestos dust from James Hardie. I saw it with my own eyes. I can only hope that none of them ever experience what I'm going through.'[11]

Like Cini, Lawrence Buttigieg, diagnosed with mesothelioma at forty-seven, had helped his parents build and maintain their market garden. 'Dad found out Hardie was giving away for free asbestos offcuts and tailings which many other Maltese people were using in driveways,' he said. 'I would get onto the back of the truck, shovel and rake the asbestos ... onto the ground ... The asbestos fibres were very fluffy, much like very soft soil and sawdust, but moved around more like sand.'[12]

Most of these market gardens are now gone, bulldozed away with the rapid development boom in western Sydney after 2000. No warnings were sounded about the lethal dust from their pathways. There is no evidence that government inspections were made of the asbestos-contaminated lands. Certainly Hardie never advised the public of the potentially deadly hazard as tonnes of its waste were scraped away and cleared.

In the meantime, in every capital city where the company had its factories lurk the anonymous killer driveways. Most are showing signs of wear now, with cracks developing and the asbestos more friable. Some owners, oblivious to the risk, break them up with sledgehammers and picks, the dust flying.

I asked Hellicar if she was concerned about the possible danger to the public, especially given that children could be exposed. She was quick to answer:

If you're saying, 'Should James Hardie pay for a clear up?' No, at the end of the day. And why 'No'? Because we cannot be a bottomless pit. The fact of the matter is ... this was not some James Hardie conspiracy to foist a product on the world. Governments were there, companies were there. We all were party to this great new product and at some point we have to just all recognise there was a big mistake made about asbestos.[13]

Hellicar knew the stakes. For the previous decade, she had received a stream of reports to the Hardie board warning of the 'catastrophic' legal liability if the company had to remediate contaminated sites. In fact, in 1999, seven years after she'd joined the board, Hardie experienced a nasty scare when householders in a suburb of Auckland realised that thousands of tonnes of asbestos waste from the Hardie factory at Penrose had been dumped into a gully at the edge of their blocks.[14] It had been farmland, but was later subdivided by Manukau City Council and sold as residential lots.

Cliff McCord, one of the Flatbush residents who bought his block in 1994, told me you could scoop asbestos waste up by the bucket-load in the back garden.

'We'd just ignored it. It was all broken up ... bits of pipes and board. We just took it to be builder's fill,' he said.[15]

Council tests revealed that it was mainly blue and brown asbestos. A protracted four-year legal battle broke out between the council and the landowners; eventually, the council removed about 6000 tonnes of contaminated soil at a cost of more than $2 million.[16]

In Australia, Hardie's lawyers and advisers anxiously watched the New Zealand dispute, fearing that the responsibility might be sheeted home to the source of the waste.

During April 1999 Hardie's PR adviser 'LJ' Loch monitored the dispute on an almost daily basis, reporting to her boss Greg Baxter and Hardie's in-house lawyer, Wayne Attrill: 'The issue does not appear to have gained media prominence today ... Manukau Council has been left high and dry by the Health Department which announced last night that federal funding would not be forthcoming for further testing ...'[17]

Loch quoted various reports in the media, including an account by a contractor, Fred Thomas, who told a local radio show that he'd delivered the waste to farmers for thirty shillings a load.

'It was in very big demand by the farmers,' he said. 'Some of the employees at Hardie's used to get me to deliver it to their house ...'[18]

Loch's concern was not simply that someone might sue Hardie. Since the company had switched to asbestos-free fibre cement in 1987, the worry now was that the adverse publicity could affect sales of its modern product.

There have been a lot of references to James Hardie, Hardie products, plus JH brand names which have significant potential to impact on public perceptions of fibre cement. I am currently working on a strategy to deal with this and NZ is currently exploring an excellent opportunity to complete a house for a disabled Maori woman who had spent her life savings on a house which was three-quarters complete when the builder went into liquidation. If we went ahead with this it would be an opportunity to reinforce the difference between fibre cement and fibro cement, i.e., NO ASBESTOS.

The draft media strategy Loch drew up the following month in conjunction with Baxter, Attrill and CEO Macdonald observed that there had been '... unfounded comments made on talkback radio alleging unethical behaviour by James Hardie in unnecessarily and wrongfully exposing workers and the public to asbestos'.[19]

A small survey of public attitudes commissioned by Hardie revealed 'a perception that James Hardie was not environmentally friendly or conscious', even though New Zealanders did not appear to have directly linked the Manukau controversy to Hardie's local operation. Loch recommended the company continue to monitor the media attention and take steps '... to enable James Hardie to anticipate, manage and neutralise outbreaks of community concern such that impact on our sales or reputation is minimised'.

She also flagged possible problems for Hardie's local staff in Auckland who '... may find themselves under the spotlight, especially as the issue escalates'. A review of security at its New Zealand factory was suggested, because '... if attention switches from Manukau City Council to the manufacturers, there may be a public perception that we have taken advantage of a "legal loophole"'.

Just as Hardie began planning its asbestos separation strategy in February 2000, the company's directors were informed by its general counsel Peter Shafron that the New Zealand threat had worsened. Legislation had been proposed to impose retrospective liability on polluters to remediate 'historical contaminated sites'. There was a risk that Hardie could be expected to foot the bill of cleaning up its waste.[20]

The board resolved to 'commence a campaign to oppose or influence the proposed amendments' — a campaign which apparently succeeded. Although the details of its lobbying have never emerged, plans were drawn up to brief key New Zealand politicians, and under a new Labor government the proposed legislative changes never eventuated.

Hardie's New Zealand experience was mirrored wherever the company had operated. Apart from the driveways and paths that employees and others were encouraged to construct, the company had also deposited bulk asbestos waste in a multitude of locations across the country.

I first became aware of the practice in 1978, when the former Hardie engineer, Fred Sandilands, contacted me after my ABC radio series about the industry had aired and I helped publicise his story.

Sandilands was forty-nine and had worked at Hardie for most of his life. He had remarried in 1975 and left the company to start a new life in Singapore, where he got a job with Hume, another asbestos manufacturer. After a medical checkup at Hume he was told that he had mesothelioma. He called me just after he had returned to Australia.

I had never met someone with mesothelioma before and I will never forget my interview with him.

The previous day fluid had been drained from his lungs at a nearby hospital, but the resident doctor was unfamiliar with the procedure and accidentally sank the needle in too deep, puncturing the back of the pleura. On the morning I met him, Sandilands told me he had woken before dawn. When he went to the bathroom to take his tablets, he had received the fright of his life. As he turned on the light, he said, he caught sight of himself in the mirror, and slumped against the wall, paralysed with fear. There was a lump on his back the size of a football. He slid to the floor, where he sat, sweating, waiting. The lump, it turned out, was the fluid that had escaped the pleura to gather in an enormous bubble under his skin.

He knew he was dying. He spoke calmly, but slowly, as one suffering a lot of pain. He had set his alarm to ring every four hours to remind him to take his painkillers.

Sandilands expressed disbelief that the company for which he had felt so much affection could be so tough in its compensation negotiations with him. The enormity of the Hardie cover-up was dawning on him. As his death grew closer it weighed on his mind.

His wife told me that he would sit at the window, staring into space. She would tell him to stop worrying about the compensation. 'For goodness' sake, just take whatever they offer,' she had urged.

But he would shake his head, saying, 'I'm not going to let the bastards get away with it!'[21]

Sandilands had supervised the dumping of thousands of tonnes of asbestos waste throughout the suburbs surrounding Hardie's Sydney factory. When his story went public in the newspapers and on ABC TV, Hardie's chairman, John Reid, circulated a letter to shareholders and staff because, he wrote, 'unfortunately the facts have not always been presented in full or objectively'.[22] Under the heading 'Setting the Record Straight',

Hardie set out a response carefully crafted by its PR consultant, Bill Frew: 'Because the small amount of asbestos fibres in our products is *locked in* by cement, it cannot escape into the atmosphere as dust, and therefore poses NO RISK TO HEALTH.'

And what mention did the Hardie's statement make of its dumped asbestos waste?

> Disposal of waste, which consists mainly of off-cuts from the finished product and wet waste consisting of ninety per cent cement, sand, etc., and a small amount of asbestos, has been carried out in accordance with Government imposed regulations applicable in each State. The NSW Health Commission is now investigating two sites in the Parramatta region where waste asbestos cement has been dumped, but a Commission officer has already been reported as saying there did not seem to be a health hazard.

Once again, the carefully chosen words did not actually say the waste was safe: an officer was 'reported' as saying there did not 'seem' to be a danger.

Behind the scenes, both Hardie and the Health Commission were scrambling for cover. There were many more sites than the two mentioned. Frank Stewart, the NSW health minister, urged householders not to be alarmed.

'Asbestos dust does pose a health hazard, but it requires exposure over a long period of time,' he said soothingly and quite inaccurately.[23]

The government soon identified dump sites at other Sydney suburbs, among them North Rocks, Wentworthville, Granville,

Silverwater, Homebush and Parramatta Park. Parramatta Council identified several areas, including a public reserve at the end of Ruse Street in the centre of the satellite city, where 'substantial quantities' of asbestos had been dumped.[24] James Hardie quietly offered the council $10,000 as part-payment for the cost of covering it with topsoil.[25]

Senior executives at Hardies had been concerned about waste disposal for some time before this. Sandilands had met regularly with Dr McCullagh and others on the company's Environment Control Committee since 1971.[26] At their first meeting they decided to monitor the asbestos dust levels wherever Hardie waste was being dumped. The minutes noted: 'it was agreed this should be done — but with caution.'[27]

Around the country, the 'cautious' monitoring began. In Perth, waste was then being trucked as landfill onto the grounds of a Christian Brothers orphanage at Castledare, where a model railway was to be built near the river as a tourist attraction. Tony Abela was one of the orphans who helped spread the waste.

> There were probably eighty boys involved, from age eight up
> to ten … We saw these big trucks coming through … the
> Brothers got the young boys with shovels into the blue stuff,
> the asbestos, digging it and spreading it. It was blue in
> colour, like a very fine, powdery substance … They got their
> shirts full of asbestos, to dump it in different areas.[28]

Hardie dispatched its industrial hygienist, John Winters, to discreetly test the dust next to the orphanage. The results were alarming. Downwind of the tipping he measured 280 fibres per cubic centimetre — over sixty times the level then deemed 'safe' for an eight-hour shift in the Hardie factories, a level soon to be halved. And for blue asbestos, the company had long realised that there should be zero exposure.

Hardie's minutes recorded that '... Mr Winters questioned the wisdom of tipping in the grounds of an orphanage and urged that more attention be given to the prompt covering of tipped asbestos waste'.[29]

Thirty years later, Hardie waste in Perth was still being dug up in the road and rail reserves at Burswood where the company had dumped it from its nearby factory. Nearly three-quarters of the surface ground samples taken in 2001 showed the presence of blue, brown and white asbestos.[30] Other landfill excavated from Goodwood Parade in Riverdale to help construct a new freeway was discovered to be 'massively contaminated'.[31]

West Australian health authorities expressed surprise, but played down any possible dangers, expressing satisfaction that test results had demonstrated 'no risk to public health'.[32]

In 2003, when I interviewed Hardie's CEO Peter Macdonald about its near-bankrupt foundation, he repeatedly asserted that the company had always done 'the right and proper thing'. Tiring of the PR mantra, I switched the subject mid-interview. 'What about the waste?' I asked. 'Has Hardie ever itemised actually where it put the waste and made sure that there isn't any more lying around?'

'James Hardie has always done the right and proper thing,' came Macdonald's unflappable reply, as the PR chief Greg Baxter suddenly began to make noises indicating it was time to wind up the interview. The CEO continued: 'So where materials needed to be disposed of, historically, then it's always got appropriate permits and permissions to locate its materials.'[33]

This was really a half-truth. Federal and state government authorities had failed to regulate the safe disposal of asbestos until the late 1970s. Hardie had been able to exploit this omission. The company had noted in 1972 that: '... before long, state or local government authorities would maintain a register of those tips at which asbestos-containing wastes might be dumped.'[34]

The practice was the same throughout the country. In Queensland hundreds of tonnes of the waste were used to fill in swampland around the Meeandah area.[35] In South Australia, when the young John Reid had been manager during the first half of the 1960s, the practice was rife, and continued long after he moved to head office.

Terry Miller, who worked for Hardie for nearly twenty years at its South Australian pipe factory in the Adelaide suburb of Elizabeth until its closure in 1987, later developed asbestosis. His wife, who used to wash his dusty overalls, died of lung cancer, although at the time nobody thought to investigate the possible asbestos connection to her death.

Miller and his workmates used to be covered in dust. In his court papers he told of one occasion when they went to the local hotel for lunch: 'As we walked into the pub we looked behind us and we could see the trail of dust we had left behind ... we laughed.'[36]

Across the road from the Hardie factory was Pratts Caravans, which, said Miller, 'constantly complained there were streaks of greyish dust on their white caravans'. As in Sydney and elsewhere, the public used to take trailer loads of the scrap asbestos to make paths and driveways. According to Miller, '... James Hardie asked for a donation of ten shillings a trailer load, which went to a local charity'.

One of Miller's jobs was to drive tonnes of asbestos waste to the local dump. During the early 1970s he dropped the waste at the corner of Adelaide's Heaslip and Womma roads; he told me that in later years he drove truckloads to the city's St Kilda tip, where the sea breeze used to whip up the dust. 'When I dumped my load, dust, pieces of pipe and chips went everywhere. If the public were in the wrong place at the wrong time, they'd get covered in the stuff. At times you couldn't even see them when you tipped the stuff out.'[37]

Hardie knew the public was at risk. Its secret monitoring in Adelaide during November 1974 revealed that even twenty minutes after tipping had stopped there was an asbestos dust count of thirty fibres per cubic centimetre. Hardie's nominal 'safe' level of exposure for a worker during a shift in its factories at the time was four fibres per cubic centimetre, soon to be halved. Hardie's Dust Committee warned of '... the harm that such an event could cause to the Company's good public relations' if the news leaked out.[38]

But the public remained blissfully ignorant. It had no idea at the time that the dust was carcinogenic, nor that the levels at the community tip were at least seven times higher than the already unsafe standard applied in the Hardie factories.

Decades later, the public danger persisted. In 2004, Terry Miller showed me raw asbestos waste still close to the surface at the reclaimed tip at St Kilda. The remains of large asbestos pipes, some badly frayed, were being used to hold the soil near a boat ramp in the public reserve.

The question lingers: how do the former James Hardie executives and board members justify their continuing silence about such hazards? These are people who prided themselves on their sense of civic responsibility.

One former senior manager at the Sydney head office only spoke to me on the condition he remained anonymous. He gave me a unique insight into Hardie's thinking.

For many years, he supervised the company's waste-disposal practices, during which time many of his former colleagues died from asbestos disease. His loyalty to Hardie was unshakeable, even though he knew that the company had once endeavoured to conceal the asbestos scarring in his own lungs.

These days, he told me, wherever he went he kept an eye out for the asbestos driveways that he knew were so popular among his fellow employees.

'I deliberately look,' he said. 'Particularly with different friends that I knew had done it.'

What loomed larger in his mind, he told me, were the areas around Sydney's Parramatta region where the company had dumped thousands of tonnes of asbestos waste.

The classic example was the Rosehill Bowling Club. They built that hotel next door to the club and as soon as they started digging (I actually went over there), I seen it. They'll never be able to do anything about that, because that is probably twenty feet deep! And all different factories along the river ... it surprises me. You come up in the [Parramatta River] ferry — have you done that? It's all poking out of the banks! I did it on purpose about a month ago. Somebody had told me, so I went and got the

ferry ... and there it was! And they were saying to me,
'Well, you tipped that there!'

I *had* done the trip myself. At low tide, the asbestos was clearly
visible among the mangrove roots. Did he worry about the
potential danger to builders who might excavate sites where it
had been dumped? 'They'll know straight away once they start
digging there,' was his unconvincing reply. 'As soon as they dig
it up, they'll see it, won't they? They'll know what it is.'[39]

At Greystanes near Parramatta he'd tipped 'tonnes and
tonnes', he told me.

> Once I started to worry about it, I went out to make sure it
> was alright. What's happened there is really great, because
> they've got a football field there and they have put six feet of
> dirt over it and it's ideal, because of the drainage. I know of
> a lot of areas that I have filled that I have gone and looked
> at, and I now know they're alright.'

On one occasion he called the former chairman, John Reid,
about a dump site that needed remediation.

> There was one here recently where I knew we had tipped it.
> These people rang up, and they'd started to dig and they
> said, 'Look, do you remember this?' And I said, 'Yes, I put it
> there!' And the company [James Hardie] paid the $200,000
> to get it out. No argument or discussion on it whatsoever. I
> was able to go back and say, 'Yes, I did that'. I went to JB
> [John Boyd Reid]. 'No argument over that,' he said. 'We'll
> do it.' And they done it!

For the former executive, this action by Reid was proof of Hardie's bona fides as an honourable company. But there is another view. If the story were true, then here was Reid, a man who lectured budding company directors on ethical corporate behaviour, still organising money to quietly remediate a site that his family company had polluted with carcinogenic dust. Why the secrecy? And what about the people at other dump sites? Why not alert them?

The number of people killed by Hardie products in the course of their work, whether in the company's factories or outside, was sooner or later likely to become public knowledge. So too was information about the products which had killed them. Many at the top of the firm, though, were also aware of those more invisible potential hazards, in places like the driveways and dumps, where most of those exposed would have no knowledge of its presence. Yet time and again Hardie directors and executives chose silence. One example starkly demonstrates the company's continuing failure to warn the public of a danger that still lingers, possibly in thousands of homes.

For many decades the millions of tonnes of raw asbestos shipped into Australia and transported from mines at Wittenoom, Woodsreef and Baryulgil were carried in hessian bags. Asbestos dust leaked through the bags, leaving a trail of death behind it: wharfies who had unloaded the bags from the ships; truckies who had driven them to the factory; workers who had poured their contents into the factory hoppers. Death stalked them all.

The danger became apparent to waterfront unions by the late 1960s. Hardie's Neil Gilbert reported early in 1968 a suggestion by an executive of the fellow Australian manufacturer Wunderlich:

... since Wunderlich were under so much pressure he would like to make a dramatic move which would show the unions, etc. that they were aware of the problem and were doing something constructive towards overcoming it. One such move would be to arrange for all the asbestos coming into Australia to be packaged in dust proof sacks.[40]

Gilbert rejected the proposal as 'palliative' at best, because the extra dollar or two per tonne of asbestos it would cost could be better spent controlling dust in the factories. Besides, he wrote, 'The sudden change to dust proof sacks would tend to excite the various State Health Departments and arouse them into giving us closer attention than we want at this stage ... no doubt we ultimately will be forced into using dust proof sacks, but I believe the move should come from the State Authorities.'[41]

The danger from the hessian bags did not stop at the factories. Once the asbestos had been emptied from the bags, many millions were recycled for other uses.

Cleaning the bags before reuse did not make them safe. Hardie conducted tests in December 1968 on cleaned bags that had been recycled to transport mouldings and roofing accessories.[42] When the bags were shaken, the tests demonstrated that the dust levels remained dangerous. Gilbert resolved that the bags should not be used for any further purpose. His instruction was ignored, but in any case it was already too late for many unsuspecting victims.[43]

John Harris, who worked as a Hardie sales representative in its Newstead factory in Brisbane, remembered that a contractor called the office regularly to arrange to pick up the bags for recycling.[44] They were popular among banana farmers to drape over the fruit as protection from birds and bats.

According to Jim Dobson, a former chairman of the Banana Growers' Council, until the early 1970s hundreds of growers used them. 'That must have been when they woke up to the dangers, but it was kept a very close secret even then that there were problems with it,' he told a reporter. 'We used to get them in bundles of twenty. We cut them down the side and hung them around the bunch. They were very effective.'[45]

Decades later, the banana growers began to die.

At first Fred Bowden, who had grown bananas in southeast Queensland, was told he had asthma. His condition worsened, until he was diagnosed with mesothelioma. For the forty-eight-year-old father of three, the news came as a complete shock.

'Some of the bags had six inches of asbestos in the bottom,' he recalled.

Paul Paroz, another son of a banana grower, developed pleural plaques from his exposure to the bags as a child; in 1998 Hardie settled his case out of court. The company had hired private investigators to spy on him before the trial date, but its lawyer reported that '... surveillance yielded no useful information as Mr Paroz went into hiding before the trial'.[46]

In Victoria, a similar picture began to emerge. Bag merchants had regularly called in at the James Hardie plant in the Melbourne suburb of Brooklyn. Many of the asbestos bags were then reused by market gardeners in Victoria Market to carry vegetables and fruit. Once again, a string of unsuspecting farmers and fruit and vegetable sellers were to die horrible deaths as a result.

A physics teacher, Dr Tony Nash, was only five years old when he helped fold the hessian bags his father used to transport vegetables to the market. More than forty years later, in 1998, he contracted mesothelioma. Nash retained the plaintiff lawyers Slater & Gordon, who ran newspaper advertisements seeking information about the bags. They received a stream of information from the public. Many callers remembered their use in Victoria Market to carry potatoes, pumpkins and other produce — even rabbits! Customers, too, were given the bags to cart away their purchases. It seemed likely a number of market sellers had died from undiagnosed asbestos-related diseases. Nash's father, a non-smoker, had died from lung cancer when Nash was thirteen. His mother died from oesophageal cancer. James Hardie reached an out-of-court settlement with Nash.[47]

The hessian bags were put to other uses potentially even more dangerous. Robert Vojakovic of the Asbestos Diseases Society of Australia has long been worried about them. In Wittenoom, where he worked briefly, they were used to transport the mine's deadly blue asbestos to the Hardie factories in Perth.

A chilling photograph reveals laughing children hopping along a road lined with tailings at Wittenoom during the annual sack race, clutching the asbestos bags around their waists. Several of the children contracted mesotheliomas in later life, although their exposure to asbestos dust in the mining town was such that it was impossible to distinguish whether the bags or the general levels of dust in Wittenoom's roads, ovals and gardens were to blame.

West Australian wheat farmers received their superphosphate fertiliser in recycled asbestos bags. In 2006, Professor Bill Musk at the Sir Charles Gairdner Hospital examined a sixty-year-old woman who had calcified pleural plaques in her lungs, a sure sign of asbestos exposure.[48] Close questioning established that she had helped her father shake the recycled bags before returning them to their wheat farm's fertiliser supplier. Her father survived until he was ninety, but her mother was not so lucky, dying from mesothelioma at age sixty-nine. Her brothers also had pleural plaques. Musk sounded a warning that many farmers could have had similar exposure because the use of the bags in the fertiliser industry was 'endemic' from 1943 to 1966.

In Perth two former employees of the Fremantle Bag Company contacted Vojakovic. Both men were dying from mesothelioma. Kevin Barrett, who was seventeen years old when he began work at the bag company in 1969, told him that the bags had been picked up from the Hardie factories for recycling. Those which could be repaired were put into a vibrating machine to shake off the asbestos, a process that released clouds of dust. They were then reused at the Hardie factories, local vegetable markets and elsewhere. He remembered that the dust would cake in his hair and on his work clothes. The damaged bags, he said, were pulped and used to make carpet underfelt or as fill for mattresses.[49]

The revelation that the bags had been used in this way alarmed Vojakovic, who in early 2006 issued a warning to the public: 'People should be very wary when renovating old homes with original carpet and if they come across old mattresses. It is not only in Western Australia, it is throughout Australia. The recycled bags were available in every state.'[50]

What particularly concerned Vojakovic was the prospect that fibres still lodged in the underlay would be released when the carpet was disturbed. Even walking over it, he thought, might dislodge the microscopic particles; certainly if the carpet were ripped up for replacement, there was likely to be dust released.

The scale of the potential hazard was mind-boggling. Just Wittenoom alone produced 160,000 tonnes of asbestos, enough to require several million bags each containing around 45 kilograms of the raw asbestos. During more than half a century of Hardie asbestos production, many hundreds of millions of bags were used. It is impossible to discover how many homes might have the suspect underfelt, or where they might be.

The Western Australian health authorities played down the danger.

'Our investigation has found that the majority of bags that were recycled to make carpet underlay had been used to transport wheat and wool,' said Jim Dodds, the director of environmental health. 'It is believed that only one in twenty-five to thirty bags used to transport asbestos was recycled. It is most likely that most of the asbestos fibres would have been removed from the bags during the cleaning and processing stage.'[51]

Yet Hardie's own tests on cleaned bags in 1968 had revealed dangerous dust levels. Nobody was suggesting that *all* carpet underlay was made from recycled asbestos bags, but it was certainly a substantial amount. Even using the WA health department's figure that only one-thirtieth of the asbestos bags were recycled, at a best guess, there would have been many millions. In an effort to assuage public fears, the department announced that it had sent samples of old carpet and underlay to the National Association of Testing Authorities (NATA), which had confirmed 'there were no asbestos fibres present'.

When I called the department for more information, an official told me that 'six or seven samples of carpet underlay from that era' were sent to NATA. 'The tests showed no evidence whatsoever of asbestos,' Mike Lindsay reassured me. 'We issued the media statement to provide some level of assurance to the public.'[52]

I asked him how statistically significant was a sample of six out of tens of millions, particularly when there was no certainty that the carpet underlay tested was even made from bags used to transport asbestos.

He agreed that '… the number is fairly low … but I don't think it's a huge risk.'

Lindsay was surprised to learn from me that at least two employees of the Fremantle Bag Company had already died from the disease. I was equally surprised that he didn't know. The bag company seemed the obvious place for the department to check.

Vojakovic said the department's carpet tests were quite meaningless. To properly ensure people's safety, those involved in the removal of old carpets should be warned to take appropriate precautions, such as testing a sample of the underfelt before its removal. He was furious with the department's response. 'To say it's a small risk is not good enough. It's complacency. People should still be careful. Any fibrous underlay is a risk and anyone who finds it should tread with caution.'[53]

Back on the east coast of Australia, in Sydney, Hardie's production headquarters, the potential hazard could be greater.

John Downes worked for the Active Bag Company based near Sydney's Mascot airport for around three years, until 1965. Downes, who later developed asbestosis, remembered picking up hessian bags from Hardie's factory at Camellia. His company sent two trucks over to the factory every week, each of which returned with two bales on the back containing between 800 and 1500 hessian bags. After the bags were tumbled in a machine to remove the obvious raw asbestos, Downes said, they were sold to various firms for use as 'carpet underlay or onion bags'.[54] Just this company alone would have processed around a quarter of a million asbestos bags every year.

The revelation that asbestos fibres lurked under Australian carpets was no surprise to Hardie. The company has known of the hazard for over thirty years, but done nothing, as the private comments of one of its public spokesmen clearly reveal.

Hardie's corporate relations manager Ray Palfreyman, whom I interviewed for the ABC radio series in 1977, had been manager of the company's Welshpool factory in Western Australia at the time when the Fremantle Bag Company still picked up the hessian bags which would later kill its employees.

During my interview Palfreyman suggested that alarmist stories by journalists about asbestos were the real problem for a 'responsible' company like James Hardie, which had done all a company could do to ensure the safety of its product.

Yet Hardie's files reveal that only six years before the interview, on a visit to the company's Victorian factory at Brooklyn, Palfreyman had spotted a pile of the empty hessian bags, which he was told were awaiting collection '... to be pulped for the manufacture of carpet underlay'.[55] The implications of this discovery must have been obvious to him. Throughout the country, the pulped underlay with its deadly fibres was lurking beneath the feet of thousands of householders. When, in time, these carpets were ripped up and replaced, dust was likely to be released. There was no indication that Palfreyman, nor any other member of James Hardie's

Environment Committee at the time, ever felt obliged to warn anyone about the carpet underlay. The committee's minutes simply record that '... it was thought the practice would not continue for long, since less and less asbestos was being received in jute bags'.

Out of sight, out of mind.

Meredith Hellicar echoed Palfreyman's refrain thirty years later: the people at Hardie had always done their best to set the highest of ethical standards, she suggested. The fact that the company's product had killed people was a 'big mistake', one for which the government should also share responsibility.

Indeed, she told me in the wake of the scandal that surrounded the company after the Jackson Commission, Hardie's former chairman John Reid was 'heartbroken at the perceptions of him, completely and utterly heartbroken'. Reid was a gentleman, 'a very proper, thoughtful, kindly, considerate man', a philanthropist who could be relied upon to discreetly donate some of his considerable wealth to sponsor worthy causes.[56] Reid himself professed to adhere to the highest standards of corporate behaviour.

Yet secrecy dominated Hardie's decisions about the asbestos health risk. Long past the time that Reid and his colleagues learnt of the dangers, workers, their families and mere passers-by continued to be exposed to potentially deadly doses of asbestos. Some still are. Rather than warn them Hardie sought to conceal the scale of the tragedy. The less said about the dangers, the fewer claims were likely. The prospect of having to disclose the true scale of people who could be killed by the company's asbestos had driven Hardie's separation from its asbestos companies. Those responsible, the modern Hardie executives and advisers, were drawn from the cream of Australian business; they also ignored the moral dimensions of their decision to cut the asbestos liabilities adrift.

In his 2004 book on corporate governance, Reid concluded: '... there is a moral dimension and imperative to all human activities. Business is no exception.'[37]

Unless, of course, the business was James Hardie and asbestos.

fourteen

EPILOGUE

The Australian epidemic of asbestos disease continues to grow, with alarming signs that an increasing number of victims have contracted disease through exposure to dust from Hardie's products through activities such as home renovation.

Even dogs die from mesothelioma in Western Australia. The veterinarians who discovered these cancers were initially perplexed, but the Asbestos Diseases Society of Australia (ADS) in Perth soon provided an explanation. It's here that Hardie extensively marketed its Super Six fencing, which included first blue, followed by brown asbestos.

The dogs probably develop the cancer as a result of scratching and digging around the fences. The society advises callers to seal in the asbestos fibre by painting their fences until they can afford to have them replaced. Meanwhile, children bounce their balls against these fences while they play. Staff who answer the telephone at the ADS have to steel themselves before taking calls. Often those phoning have just been told they have mesothelioma, 'the fatal disease with the unpronounceable name'.

The Perth-based society has counselled more than 10,000 families and its workload is increasing. It is the most professional support group in the country — if not the world — partly because, as a registered charity, it has secure funding through the sale of lottery tickets and other activities. Most other groups rely heavily on assistance given by the major plaintiff law firms, who compete to attract new clients. These firms also provide significant sponsorship of medical research into asbestos diseases, as does the ADS.

It is also in Perth that Professor Bruce Robinson is coordinating promising research into a possible cure for mesothelioma at the Sir Charles Gairdner Hospital. His team has already developed a blood test that appears to give early warning of this cancer, and research is progressing on techniques for early intervention and treatment.

Elsewhere in the world other research offers hope. Belinda Dunne who, as a four-year-old, played next to her father in the backyard of the Adelaide family home when he broke up Hardie asbestos cement while renovating the garage, developed mesothelioma nearly thirty years later. She received experimental treatment in Pennsylvania and more than thirteen years later she is still alive and has given birth to a second child. Others with mesothelioma have lived even longer. But they are all exceptions to the rule.

Hardie's former CEO, Peter Macdonald, used the phrase 'atmospheric asbestos' to argue that Dunne's exposure may not have come from Hardie's asbestos cement. He suggested that it could also have come from a wide range of sources, including government and private buildings with loose asbestos in heating and cooling systems.

Indeed, virtually no inhabited area of Australia is free from asbestos.

Even in the remote outback, Aboriginal communities continue to agitate for the safe removal of Hardie asbestos cement sheeting used in housing construction, the result of a targeted marketing effort which continued through the 1980s. And state governments throughout the country still fund multi-million-dollar asbestos removal programs in schools.

In the US and much of Europe, the picture is similar: schools, the Capitol tunnels in Washington, and the British, Canadian and EU Houses of Parliament have or are being stripped of asbestos. In parts of Asia, by contrast, asbestos use in building construction is still increasing.

Despite the huge death toll, very few of those responsible for the production and marketing of asbestos have been prosecuted. In Brazil, France, Italy and the US, criminal charges have been pursued against executives of asbestos producers of former Eternit subsidiaries and WR Grace. Some have resulted in convictions. But in Australia, no government has ever laid criminal charges for decisions that will probably result in avoidable deaths equivalent in scale to those sustained during World War One.

On 15 February 2007, the day before the statute of limitations prevented action, Australia's corporate regulator, the Australian Securities and Investments Commission (ASIC), commenced civil penalty proceedings against those who were directors and senior executives of James Hardie Industries when it launched its asbestos foundation.

Six months later the regulator announced that no criminal charges would be laid.

Despite an 'exhaustive investigation' spanning three countries and involving 348 billion documents, 72 examinations and the issue of 284 notices to obtain evidence, ASIC said it had failed to gather sufficient evidence to support a prosecution.

In its civil case, ASIC sought penalties of up to $200,000 for each offence and a ban on the defendants from holding directorships or executive positions in public companies. Those directors targeted by ASIC who were still on the Hardie board, including the company chair Meredith Hellicar, resigned a week later. She remained a director of the giant 'wealth management' company, AMP.

When the ASIC case came before the NSW Supreme Court in late September 2008, it was conducted in the same downtown Sydney building which houses the Dust Diseases Tribunal. A few floors below was the courtroom where Bernie Banton and thousands of other asbestos victims had fought Hardie for their compensation, and where Commissioner David Jackson had conducted his inquiry into Hardie's flight offshore.

Ironically, much of the ASIC case against the Hardie defendants boiled down to a single company media release, the one issued when the asbestos foundation was launched, which claimed the foundation was 'fully funded' and provided asbestos claimants 'certainty'. These words in the release formed the core of ASIC's accusation of misleading conduct. Two former senior executives, Peter Macdonald and Peter Shafron, chose not to testify, but were represented by high-powered lawyers.

In court, none of the accused remembered who had suggested the phrase 'fully funded'. Greg Baxter, the former head of Hardie's corporate relations who then went to work for Rupert Murdoch's News Limited, admitted that he was responsible for drafting the media release, but he didn't know where this term — which now 'embarrassed' him — had originated. Ultimately, he said, more senior managers were responsible for content of the release. Nor did any of the accused Hardie directors admit to approving the release, despite the minutes of the board recording their assent. Some directors denied it outright; others could not remember.

It emerged in court that, when the company was later locked in negotiations with the Carr government, and was seeking immunity for its directors to be included in any settlement, Hellicar and two board colleagues had emailed declarations to their fellow directors stating that they *had* approved the media release, as revealed by the board minutes. A signed declaration from one director to that effect was admitted in evidence. Hellicar and her other co-accused later claimed these emails were inaccurate.

After proceedings that lasted six months, on 24 April 2009, Justice Ian Gzell found the company, the entire board and the three executives guilty of breaches to the *Corporations Act*. His judgement was particularly scathing about Hellicar's evidence, about which he had 'grave doubts'. Gzell suggested that the shock she displayed when confronted with a key document was 'feigned', and described Hellicar as 'a most unsatisfactory witness'. Hellicar's corporate career appeared to be at an end: within a few days she had resigned her remaining directorships.

However all directors and executives other than Peter MacDonald appealed and in 2010 the NSW Court of Appeal overturned Justice Gzell's judgements. ASIC in turn appealed to the High Court. In 2012, after one of the lengthiest and most expensive corporate prosecutions in Australia's history, the highest court in the land upheld the original judgements against them. During this time Meredith Hellicar provided a mentoring consultancy for budding directors, teaching them in particular the finer points of crisis management.

The deal for which Banton and Combet had fought so hard anticipated times of economic difficulty by building in a funding buffer, to ensure time to arrange staggered compensation payments or other measures agreed to by the NSW Government. However it was no match for the global financial crisis of 2008, which saw Hardie's US sales plummet. After pleas from asbestos support groups the NSW and federal governments provided the compensation fund with a loan facility to ensure full compensation payments continued.

Many of the powerful figures from the NSW government involved in the Hardie saga moved on to other political roles. The former premier Bob Carr registered as a lobbyist with the Rudd federal Labor government, adding the Asbestos Disease Research Foundation to his list of top corporate clients and exhibiting interest in voluntary work on behalf of asbestos victims. Kevin Rudd was replaced by Julia Gillard as prime minister, but as foreign minister sponsored a Labor Party policy for Australia to lead the world in eliminating the hazards of asbestos. In 2012 Rudd was replaced as foreign minister by Carr, who became a Senator in the federal parliament. Carr's former attorney-general Bob Debus served as a minister in the Rudd government before retiring, while Roger Wilkins, the former director-general of Carr's Cabinet Office, joined the Rudd administration as head of its Attorney-General's Department, where he remains. Greg Combet, the former union official called in to negotiate the deal with Hardie, became a senior minister in the Gillard government. And the Federated Miscellaneous Workers' Union has renamed itself 'United Voice'.

In the legal world, in 2011 James Hardie's Compensation Fund ran and lost an important test case in the High Court, which ruled that all material exposure to asbestos (including white asbestos) may be deemed a cause of mesothelioma. After 32 years John O'Meally, the country's longest serving judge, retired from the Dust Diseases Tribunal; Michael Slattery became a judge in the NSW Supreme Court. Wayne Attrill, who was in charge of defending claims against Hardie's asbestos subsidiaries, joined IMF, a company specialising in packaging law suits for a success fee of up to forty per cent of the proceeds. Michael Gill, the insurance lawyer whose ill-health prevented him answering questions about his role advising Hardie before he joined the board of its asbestos foundation, continued to practise as a senior consultant to DLA Phillips Fox and chair the Insurance Industry Code Compliance Committee.

The former Trowbridge actuary David Minty and his colleagues regrouped under the name Finity Consulting, where they continued to provide actuarial and insurance advice. Of all those who participated in Hardie's ill-fated corporate restructure, Minty was the only external adviser to suffer a formal repercussion: the Institute of Actuaries of Australia reprimanded him for two breaches of its Code of Conduct. Despite an investigation by the NSW Attorney-General's Department, no action was taken against Hardie's legal advisers at Allens.

Karen Banton remarried in 2012; Justice Kim Santow, Jeff Shaw, Neil Gilbert, Bill Mansfield, Doug Howitt and others featured in this book have since died.

And John Reid, with his honorary doctorates and award for philanthropy, remained in his harbourside penthouse, still occasionally dispensing funds to charities and advising company directors. Although he has written proudly about his time on the

James Hardie board in his book on corporate governance, he has repeatedly refused requests to answer questions about the company's asbestos business for this book. He has never appeared in court or any other public forum to explain his actions to the estimated 20,000 Australian families affected. His silence says it all.

ENDNOTES

Many of the James Hardie documents are on the public record at the NSW Dust Diseases Tribunal; others have been given to the author over the years by company insiders; still other documents were tendered in evidence before the Jackson Commission. Copies of all cited documents are in the possession of the author.

1 Burying 'An Australian Hero'

1 Banton v James Hardie & Coy Pty Limited, NSWDDT 157 1999.

2 Banton v Amaca Pty Limited, NSWDDT 7255 2007.

3 James Hardie correspondence between J. B. Reid and E. T. Pysden, personnel manager, 15–16 February 1966.

4 The company's own estimate of compensation claims is nearly 13,000, but this grossly understates the actual disease incidence. See Chapter 4.

5 *Sunday Times*, 31 October 1965, p. 1.

6 Dr S. F. McCullagh, confidential JH memo to R. Palfreyman, community relations manager, 15 September 1972.

7 Banton v Amaca Pty Limited, NSWDDT 7255 2007.

8 The 'dust milk', as it was known, had been allocated following a request by the Storemen and Packers Union, JH memo from A. R. Boswell, insulation factory manager, 30 December 1957.

9 *7.30 Report*, ABC TV, 14 June 2004.

10 At the time of writing, Kazakhstan, Russia, Zimbabwe, China and Canada are the largest asbestos producers. Its use for building products, in Asia in particular, is expanding.

11 The Australian government banned the import of asbestos effective from 31 December 2003. The maritime union had previously placed a ban on asbestos, but this was circumvented by importing the mineral through the port of Newcastle as 'chrysotile'.

12 The series was broadcast on ABC Radio National's *Broadband* during 1977 and later published as a book, *Asbestos: Work as a Health Hazard*, ABC (with Hodder & Stoughton), Sydney, 1978.

13 Interview, Tanya Segelov, 17 December 2007.

14 *Adelaide Advertiser*, 28 November 2005, p. 1.

15 Batson v Amaca Pty Limited, NSWDDT 147 2004.

16 Bernie Banton affidavit, Banton v James Hardie & Coy Pty Limited, NSWDDT 157 1999.

17 Interview, Meredith Hellicar, 22 December 2005.

18 *7.30 Report*, ABC TV, 17 August 2004.

19 Louise Eddy, 'Dust to dust', *Western Advocate*, 7 May 2007.

20 Banton v Amaca Pty Limited, NSWDDT 7255 2007.

21 According to Jan Hudson of Charles Sturt University, James Hardie's contribution 'was seen as supporting the education of young professionals for Australian business'. She offered no explanation for the abrupt disappearance of the James Hardie Scholarship, but told me that John Reid was 'an incredible benefactor and friend' to the university. Email, 29 February 2008.

22 Interview, Tanya Segelov, 17 December 2007.

23 Amaca Pty Limited v Banton, NSWCA 336 2007.

24 Interview, Jack Rush, QC, 16 December 2007.

25 Personal communication, former Hardie executive, 2008.

2 Bursting the PR Bubble

1 Chairman's Report, JH Annual General Meeting, 1971. 'The [Environmental Control] Committee has sponsored several research studies in industrial health and safety which are being undertaken in association with the University of Sydney.' Peter Russell, JH safety officer was told in 1963 by production manager Ted Heath that those at the Department of Tropical Medicine 'valued the annual grant' James Hardie gave the department, P. Russell, affidavit, Romano de Maria v James Hardie & Co. Pty Ltd, SADC, 1763 1998, p. 30.

2 M. Peacock, *Asbestos: Work as a Health Hazard*, ABC (with Hodder & Stoughton), Sydney, 1978.

3   Hancock famously told a journalist 'Somebody has to suffer for the sake of progress', when asked about the people killed by Wittenoom asbestos. Geoff Strong, 'There is much to value in the Fourth Estate', *The Age*, 24 August 2009.

3   The story of the Wittenoom mine and the struggle for justice by those it affected is best told in Ben Hills' book, *Blue Murder*, Sun Books, Sydney, 1989.

4   JH confidential memo, 'ABC Science Program', Dr S. F. McCullagh, chief medical officer to general manager, environment, 8 July 1977.

5   'Biological effects of asbestos', *Annals of the New York Academy of Sciences*, vol. 132, December 1965.

6   British Occupational Hygiene Society Sub-committee on Asbestos, 'Hygiene standards for chrysotile asbestos dust', *Annals of Occupational Hygiene*, no. 11, 1968, pp. 47–69.

7   H. C. Lewinsohn, 'The medical surveillance of asbestos workers', *Royal Society of Health Journal*, no. 92, 1972, pp. 69–77.

8   Selikoff to Lewinsohn, 13 August 1973, Turner & Newall Archive, no. 72, pp. 651–666. See also G. Tweedale, *Magic Mineral to Killer Dust*, Oxford University Press, 2000, pp. 226–231; and B. I. Castleman, *Asbestos: Medical and Legal Aspects* 5th edition, Aspen Publishers, New York, 2005, pp. 274–281, about how the industry tried to suppress this information and gag Selikoff.

9   B. I. Castleman, ibid., p. 274.

10 P. Brodeur, 'Reporter at large', *The New Yorker*, 12 October 1968; *Expendable Americans*, Viking Press, New York, 1974. Brodeur subsequently wrote *Outrageous Misconduct: the Asbestos Industry on Trial*, Pantheon, New York, 1985.

11 Interview, Paul Brodeur, 1977.

12 Timothy Hall, 'Is this killer in your house?', *The Bulletin*, 6 July 1974.

13 JH memo from director Frank Page to E. T. Pysden, 'New York Academy of Science', 22 March 1966.

14 JH memo, Frank Page to Ron Bolton, 6 June 1975.

15 J. B. Reid, JH AGM, 1976.

16 M. Peacock, op. cit., p. 60.

17 JH Environmental Control Committee minutes, 16 February 1977, pp. 12–13.

18 David Jackson, *Report of the Special Commission of Inquiry into the Medical Research and Compensation Foundation*, NSW Department of the Premier and Cabinet, NSW Government, September 2004, p. 59.

19 H. A. Hudson, manager, JH Factory Operations, memo to factory managers, 2 March 1976.

20 JH report on Victorian Factory Managers' Conference, 11 July 1966, p. 15.

21 JH Environmental Control Committee minutes, 18 August 1977, p. 26.

22 ibid., p. 25.

23 Confidential minutes of JH meeting no. 5, 'Dustless Methods of Cutting Asbestos Cement', 17 October 1977.

24 Mark Colvin and Nick Franklin, then ABC Radio news cadets, packaged the material for *2JJ*.

25 Letter from Barry McCrea, group operations manager, International Public Relations, to the director of the ABC Science Unit, 27 July 1977.

26 The Fraser government secretly advanced a loan of $1.4 million to the Woodsreef Mine operator, according to Cabinet documents made public on 1 January 2009, National Archives, Cabinet decision no. 7004, 26 October 1978.

27 Industries Assistance Commission report, *Asbestos*, 30 October 1979, AGPS, Canberra; 'Abandoned Asbestos Mine Causes Community Outrage', *7.30 Report*, ABC TV, 13 August 2008.

28 Letter from Leslie Anderson, Eric White's NSW director to David Macfarlane, JH managing director, 31 May 1978: 'Hill & Knowlton, our parent company, has been very much involved in this whole matter. In the UK, our office was instrumental in the establishment of the Asbestos Information Committee in 1969 ... In the US Hill & Knowlton began work for Johns-Manville in

1967 and in 1971 ... began work for the Asbestos Information Association.' Reid Collection, State Library of NSW.

29 Draft memo to JH state managers from Eric White's Bill Frew, 13 July 1978, shortly after the PR agency had been hired by Hardie over the asbestos issue, Reid Collection, State Library of NSW.

30 Appendix to the minutes of JH Environmental Control Committee meeting, 1978.

31 The minutes of the JH Environmental Control Committee record that Hardie's Ray Palfreyman told a meeting of the International Asbestos Information Conference that with the importation of such products into Australia '... the industry's position in Australia could be pre-empted'.

32 *Choice*, Australian Consumers Association, September 1977, and again in July 1978.

33 JH Environmental Control Committee minutes, 'Climate of Opinion', 4 November 1977, p. 29.

34 Handwritten notes by Ron Bolton, JH community relations manager, record a hurriedly convened meeting between Hardie and Eric White the day before the AGM, 'Main points following discussion with Eric White re Annual General Meeting Tuesday'. The minutes also note 'Consider Peacock unlikely to attend ...', 25 July 1978, Reid Collection, State Library of NSW.

35 Deidre Macken, 'Hardie parries health protest', *The Australian*, 28 July 1978.

36 'Hardie faces barrage on asbestos diseases', *Daily Telegraph*, 28 July 1978.

37 Ron Bolton's handwritten notes, op. cit., Point 6: 'Labelling'.

38 Letter from Eric White's Leslie Anderson to JH's Ron Bolton, 1 August 1978, Reid Collection, State Library of NSW.

39 JH document headed 'Pipe Sales, Review of 1981/2', Paragraph 6.2, 'The Environment', Reid Collection, State Library of NSW.

40 Conference report of meeting between Coudrey Dailey Pty Ltd and JH's Ron Bolton and others, 'Consumer & Builder Research', 1 May 1979, Reid Collection, State Library of NSW.

41 JH Environmental Control Committee minutes, 'Associations &
    Committees', 6 March 1979, p. 36.

42 Max Austin, executive director of the largely JH-funded South
    Pacific Asbestos Association, quoted by Robin Osborne, 'Cutting
    the asbestos risk', *Far Eastern Economic Review*, 22 August 1980.

43 'Public Relations Action Plan for James Hardie Asbestos', Eric
    White & Associates, September 1978, p. 12, Reid Collection,
    State Library of NSW. The 'extreme positions' taken by NSW
    were ultimately adopted by all states at the federal NH&MRC
    meeting in December 1981.

44 JH notes in the job description for community relations manager,
    'Competitive/Technical Situation', 1980, Reid Collection, State
    Library of NSW.

45 *Nationwide*, ABC TV, 15 March 1979.

46 Roger Franklin, '1000 homes on "suspect" health checks', *Sun-
    Herald*, 11 March 1979.

3 The Insiders

1   Peter Russell, personal communications, 1977, and subsequent
    interviews 2004–2006.

2   *Fibrolite Trimmings* (Hardie's in-house magazine) vol. 1, no. 1, p.
    5, Reid Collection, State Library of NSW.

3   JH memo from A. D. Woodford, 'Asbestosis', 21 February 1962.

4   Rafferty v James Hardie & Co Pty Ltd and Wallace Henry
    McCulloch, SASC, 19 March 1991, p. 234.

5   Some other Hardie managers such as Gilbert denied the existence of
    a 'Dust File', but at least one other executive confirmed its
    existence, and given Russell's acknowledged credibility there seems
    no reason to doubt him.

6   JH memo from A. D. Woodford, op. cit.

7   Russell's handwritten list is in the author's personal possession.

8   E. R. A. Merewether and C. W. Price, 'Report on Effects of
    Asbestos Dust on the Lungs & Dust Suppression in the Asbestos
    Industry', HMSO, London, 1930.

9   Report by WA factory inspector R. G. Mooney forwarded to the
     Commissioner for Public Health, 24 May 1935; see also State of
     WA v Watson, Australian Tort Reports, pp. 80–226.

10   Several former executives claim the restructure was made to
     prevent the company from having to go public. But Ian Mutton, a
     legal consultant for Hardie's competitor CSR, observed to the
     author that although this Hardie corporate restructure may have
     been made for other reasons '… it seems really odd that at a time
     when they were first put on notice that there was a lot of ill health
     flowing from the factory, they sorted the company out'. Interview,
     Ian Mutton, 5 July 2005.

11   Jones v James Hardie & Co Pty Ltd, Workers' Compensation
     Commission of NSW, no. 275 of 1939, p. 129.

12   R. Doll, 'Mortality from lung cancer in asbestos workers', *British
     Journal of Industrial Medicine*, vol. 12, 1955, pp. 81–86. See also
     B. I. Castleman, op. cit., pp. 81–89, for an account of the
     industry's efforts to suppress and alter this information.

13   J. Wagner, C. A. Sleggs & P. Marchand, 'Diffuse pleural
     mesothelioma and asbestos exposure in the North-West Cape
     Province', *British Journal of Industrial Medicine*, vol. 17, no. 4,
     1960, pp. 260–271.

14   J. C. McNulty, 'Malignant pleural mesothelioma in an asbestos
     worker', *Medical Journal of Australia*, vol. 2, 15 December 1962,
     pp. 953–954.

15   JH memo from P. Russell, safety engineer and fire officer, 'Factory
     Air Pollution', 26 April 1961, p. 2.

16   P. Russell, affidavit, Romano de Maria v James Hardie & Co Pty
     Ltd, SADC 1763 1998, p. 30.

17   Memo from Dr R. Gillott, industrial medical officer, Snowy
     Mountains Hydro-electric Authority, 23 March 1954, and
     associated documents, including inspection report from H. M.
     Whaite, scientific assistant, NSW Department of Health, Division
     of Industrial Hygiene, 9 April 1954, National Archives of
     Australia, Series A12943, Item IB90A.

18 Confidential JH memo from Dr S. F. McCullagh, 29 March 1966, Annual Medical Report 1968, Findings of the Medical Review 1969.

19 Confidential JH memo from P. Russell, 10 November 1964.

20 Interview, Ron Hinton, 15 September 2008.

21 Interview, Warwick Lane, 7 October 2005.

22 Interview, Neil Gilbert, 13 July 2005.

23 JH memo from N. Gilbert to A. Hooper, 'Proposed Asbestos Handling System', 16 November 1954.

24 As a gesture of appreciation for his work, Hardie's deputy managing director, Jock Reid, provided Gandevia with introductions to UK industry figures and money for a study tour of Britain. JH letter from N. Gilbert to Dr Gandevia, 7 August 1963.

25 See report of Ted Heath's visit to the US, JH Factory Manager's Conference 1947.

26 Brian Carroll, *A Very Good Business*, James Hardie Industries, Sydney, 1987, p. 134.

27 Interview, Warwick Lane, op. cit.

28 JH memo from F. Page to N. Gilbert, 'Dust Control', 21 April 1966. Page wrote: 'The control of dust in the Camellia area must be regarded as urgent and important. It is not possible nor is it, I feel, desirable to try and define a safe concentration. All dust is harmful ...'

29 From 1947, James Hardie had an agreement to exchange technical information with the US-based Johns-Manville relating to asbestos production; by 1951 it secured a similar agreement with the UK's Turner & Newall. Senior Hardie executives, as well as Thyne and Jock Reid, regularly met with overseas colleagues in the global cartel. Although only a few James Hardie documents have emerged to demonstrate the overseas companies shared their knowledge about asbestos health hazards, given Gilbert's ease of access in the 1960s it is inconceivable they did not. JH correspondence, Thyne Reid and Dudley Colten, Johns-Manville and Walker Shepherd, Turner & Newall, 11 October 1951.

30 Interview, Neil Gilbert, op. cit.; JH report from N. Gilbert, 'J. M. Medical Service', 1966.

31 Remarkably, no record of this very significant advice from Gilbert has ever emerged from JH discovery proceedings, but his account has never been disputed.

32 JH memo from N. Gilbert to A. D. Woodford, 'Asbestosis', 9 May 1966.

33 JH memo from Dr S. F. McCullagh to N. Gilbert, 7 July 1967.

34 JH memo from N. Gilbert to F. Page, 'Medical Review — Hardie BI', 12 December 1968.

35 Notes from JH Factory Managers' Conference, 11 July 1966, p. 15. McCullagh's warning was prophetic, but it was not until thirteen years later that NSW Health inspectors responded to a request by the Federated Rubber & Allied Trades Union to measure asbestos dust levels at the Goodyear factory which adjoined James Hardie. 'Gross contamination with fine amosite [brown] and chrysotile [white] asbestos fibres' was found; letter from A. T. Jones, Industrial Hygiene Branch, Division of Occupational Health, to M. F. Fatel, Goodyear Tyre & Rubber Co, 6 November 1979. At least one worker from an adjoining factory later contracted mesothelioma, Cassidy v James Hardie & Co Pty Limited and Melesco Properties Pty Limited and ICAL Limited, NSWDDT 142 1993.

36 JH letter from N. Gilbert to A. McDougall, 5 September 1967.

37 JH memo from F. Page to N. Gilbert, '3rd Australian Medical Congress', 23 August 1968.

4   The 'Dubious Statistics of Death'

1 JH confidential memo from Dr S. F. McCullagh to N. Gilbert, 'South Australia Health Department — Occupational Health Branch — Asbestos Survey', 18 November 1968.

2 The NSW Dust Diseases Board, as it is now known, was established as a no-fault statutory compensation body and has

its origins in the struggle for compensation for injured workers by the construction and mining unions in Sydney and Broken Hill during the 1920s. See the official history, Rosey Golds, *A History of the Dust Diseases Board, 1927–2007*, DDB, Sydney, 2008.

3 Dust Diseases Board internal document, 'Suggested letter to the Minister', responding to JH request from R. Palfreyman, manager, 11 April 1973, for the names of JH ex-employees with asbestos-related disease, August 1973.

4 D. O. Shiels' lecture 'Asbestosis', to factory inspectors, Victorian Department of Health, 11 July 1947.

5 Memo from D. O. Shiels, chief industrial hygiene officer, Victorian Department of Health to chief health officer, 're X-Ray Survey in Dusty Trades', 2 September 1953.

6 D. L. G. Thomas, 'Pneumonokoniosis in Victorian Industry', *Medical Journal of Australia*, vol. 1, no. 3, 19 January 1957, pp. 75–77.

7 JH memo, manager, Brooklyn, to N. Gilbert, 18 March 1957.

8 JH memo from A. Hooper, manager, Brooklyn, to J. T. Adamson, 6 May 1959.

9 JH letter from J. T. Reid to A. Morling, Turner Asbestos Cement Co Ltd, 18 March 1957.

10 Turner & Newall correspondence to James Hardie, January 1957, Turner & Newall Repository, Manchester, England, Drawer 115–187, File 1960 M2, Reel 24/186.

11 Report from Turner & Newall's Australian representative R. M. Stratton to A. N. Marshall, Turner & Newall, Norwich, UK, 11 September 1959, Turner & Newall Repository, ibid.

12 Turner & Newall memo from Dr J. F. Knox to J. Waddell, 'Asbestosis — Australia', 22 September 1959, Turner & Newall Repository, ibid.

13 Letters from J. W. Roberts, Turner & Newall to R. M. Stratton, Turner Asbestos Products, Australia, 11–18 September 1959, Turner & Newall Repository, ibid.

14 Report from R. M. Stratton, Turner (Aust.) to A. N. Marshall and J. W. Roberts, Turner & Newall K, 16 September 1959, Turner & Newall Repository, ibid.

15 ibid.

16 JH memo from A. Hooper to J. T. Adamson, 6 November 1959.

17 ibid.

18 Interview, Jim Donovan, 12 February 2006.

19 Letter from D. A. Ferguson, School of Public Health & Tropical Medicine, Commonwealth Department of Health, Sydney University, and acting secretary, Occupational Health Committee of the National Health & Medical Research Council, to Capt. D. M. Hogan, chairman, Federal Advisory Committee on Waterfront Accident Prevention, Sydney, 2 September 1969.

20 JH Environmental Control Committee minutes, 29 March 1971, p. 2.

21 The NH&MRC 'Recommended Code of Practice for Handling Consignments of Asbestos Fibre in Australian Ports' was circulated to branches and councillors of the Waterside Workers' Federation by T. Bull, WWF federal organiser on 3 March 1972.

22 Interview, Eva Francis, 6 August 2005. Francis provided a report to the Dust Diseases Board, 'Possible Asbestos Hazard', 8 December 1971, in which she said the wharf labourers were 'generally covered with asbestos fibre and dust as bags are often broken or damaged'.

23 Most worked for the trucking contractor York & Kerr. See, for example, Dark v York & Kerr Ltd, Telstra Corp. Ltd, Selstam Pty Ltd and James Hardie & Coy Pty Ltd, NSWDDT 98 1995.

24 Dr S. F. McCullagh, statement, McCusker v Seltsam Limited and James Hardie & Coy Pty Ltd, 26 March 1997, NSWDDT 179 1996, p. 38; report from R. M. Stratton, Turner (Aust.) to A. N. Marshall and J. W. Roberts, Turner & Newall, 16 September 1959, Turner & Newall Repository.

25 Interview, Eva Francis, op. cit.

26   NH&MRC 'Code for the handling of asbestos by small users',
     AGPS, Canberra, 1978.

27   The NSW Health Commission also attacked the NH&MRC
     document, 'The Medical Aspects of the Effects of the Inhalation of
     Asbestos', approved by the Council in October 1979, which it said
     'plays down the hazards of chrysotile to an unjustified extent' and
     'should not be issued under any circumstances'; letter from W. A.
     Crawford, director, NSW Division of Occupational Health and
     Radiation Control, 1 February 1980. Crawford had previously
     criticised the 'worker education' pamphlets prepared by Hardie as
     'misleading' for the same reason; report of meeting, 'Hazards of
     Asbestos', NSW Bureau of Environmental and Special Health
     Services, 5 October 1977. Significantly, as late as 1983 Hardie's
     Ron Bolton observed that 'My personal conviction is that our
     workforce exposed to asbestos is not sufficiently informed about
     the dangers of asbestos ...', Reid Collection, State Library of NSW.

28   Minutes, first meeting of the Asbestos Occupational Hygiene
     Committee, 11 August 1972.

29   Minutes, Asbestos Occupational Hygiene Committee, meeting 36,
     Sydney, 21 October 1977; also T. Hall, *The Ugly Face of
     Australian Business*, Harper & Row, Sydney, 1980, p. 21.

30   A. Rogers, National Occupational Health & Safety Commission,
     'The Extent of Mesothelioma in Australia and its Relevance to
     Occupational Hygiene Practice', Proceedings of the First Annual
     Conference of the Australian Institute of Occupational Hygienists,
     Ballarat, Victoria, 3 December 1986; D. Ferguson, 'Australian
     mesothelioma register' *Medical Journal of Australia*, vol. 1, 1980,
     pp. 150–152.

31   A. Rogers, ibid.

32   Letters from R. Thompson, secretary, Australian Mesothelioma
     Surveillance Program, 26 March 1987, to media and trade unions.

33   In the author's possession from the files of Dr Julian Lee,
     Australian Mesothelioma Surveillance Program, Commonwealth
     Institute of Health.

34 J. Leigh and T. Driscoll, 'Malignant mesothelioma in Australia 1945–2002', *International Journal of Occupational Environmental Health*, vol. 9, 2003, pp. 206–217.

35 Interview, William Musk, 30 May 2005.

36 G. Hillerdal, 'Mesothelioma: cases associated with non-occupational and low dose exposure', *Occupational Environmental Medicine*, vol. 56, 1999, pp. 505–513.

37 Letter from Toula Papadopoulos, OHS coordinator, Australian Chamber of Commerce and Industry, to Ms H. L'Orange, chief executive, National Occupational Health & Safety Commission, 14 February 1997.

38 Letter from Michele Patterson, assistant general manager, Occupational Health & Safety Division, WorkCover NSW, to Ms H. L'Orange, chief executive, National Occupational Health & Safety Commission, 17 March 1998.

39 Letter from Bill Frew, Eric White & Associates, to Ron Bolton, JH community relations manager, 6 February 1980.

40 KPMG, 'Valuation of asbestos-related disease liabilities of former James Hardie entities to be met by the AICD Trust', 22 May 2008, James Hardie, Sydney, p. 59.

41 Dr M. Clements, National Centre for Epidemiology and Population Health, Canberra, presentation at Accident Compensation seminar, 2 April 2007.

42 Email to M. Peacock from Julie Sheather, vice-president, public affairs, James Hardie, 11 November 2003.

43 Personal communication between M. Peacock and Samantha Stebbings, communications director, Australian Safety & Compensation Council, 24 May 2007.

44 The so-called Helsinki guidelines, later refined for Australia, 'The diagnosis and attribution of asbestos-related diseases in an Australian context', Adelaide Workshop on Asbestos-Related Diseases, *Journal of Occupational Health and Safety — Australia–New Zealand*, vol. 18, no. 5, 2000, pp. 443–452. The NSW Dust Diseases Board commissioned a paper from Alan

Rogers which strongly criticised the Helsinki and Australian meetings: A. Rogers, 'An evaluation of the exposure criteria and lung fibre burden associated with the Helsinki criteria and its applicability to Australia', Dust Diseases Board Research Report, November 2001.

45 See, for example, Restuccia v Workers' Compensation (Dust Diseases) Board, NSWDC 2172, 2173 2001.

46 D. Henderson, K. Rodelsperger, H-J. Woitowitz, J. Leigh, 'After Helsinki: A multi-disciplinary review of the relationship between asbestos exposure and lung cancer, with emphasis on studies published during 1997–2004', *Pathology*, vol. 36, no. 6, December 2004, pp. 517–550.

47 Interview, Bruce Robinson, 30 May 2005.

48 Letter from E. Francis, scientific officer, Division of Occupational Health & Radiation Control, Health Commission of NSW, to Dr Alan Bell, director, 1 March 1978.

49 Health Commission of NSW, 'Background Notes for Minister on Asbestos and Asbestos-Induced Disease', 17 April 1977.

50 KPMG, 'Valuation of asbestos-related disease liabilities arising from the manufacture or use of asbestos by the former subsidiaries of James Hardie Industries NV as at June 2004', 21 November 2004, Sydney, p. 7.

51 A ban on the importation of asbestos was imposed by federal minister Tony Abbott effective 31 December 2003. Some exemptions for use, for example by the Defence Department, remained.

52 J. C. McNulty, 'Malignant pleural mesothelioma in an asbestos worker', *Medical Journal of Australia*, vol. 15 December 1962, pp. 953–954.

53 In the document prepared by James Hardie in 1978 for its workforce, the company also emphasised so-called environmental mesothelioma: 'Not all mesothelioma is caused by asbestos, but when it is most cases are due to breathing dust containing crocidolite (blue) asbestos fibres', Asbestos

Association of Australia, 'Your Health and Asbestos', 1978. (This booklet, at Hardie's instigation, was also endorsed by Sydney University's Professor Ferguson and the general secretaries of the two asbestos unions, Ray Gietzelt of the Missos and Frank Mitchell of the AWU.)

54 M. Peacock, op. cit., p. 48.

55 Interview, William Musk, op. cit.

56 Report of Mr W. W. F. Shepherd, Turner & Newall, 'On a Visit to Australia 28 April 1951–6 June 1951', p. 8, Turner & Newall Repository, Manchester, England, Drawer 115–187, File 1960 M2, Reel 24/186.

57 ibid., p. 8.

58 Industries Assistance Commission report, 'Asbestos', 30 October 1979, AGPS, Canberra, 1979.

59 G. Major, statement, Hunter v John Meagher and Bryce Clover (trading as Meagher and Clover) and E. M. Miller Structural Co Pty Ltd and G. E. Hurst (construction) Pty Ltd and James Hardie Industries Ltd and Seltsam Ltd (formerly Wunderlich Ltd), NSWDDT 41 1990.

60 Dr S. F. McCullagh, 'Amosite as a cause of lung cancer and mesothelioma in humans', *Journal of the Society of Occupational Medicine*, vol. 30, October 1980, pp. 153–156.

61 Report of a meeting by officials from the NSW Health Commission and the Department of Industrial Relations at McKell Building, Sydney, 31 March 1980, to discuss review of recommendations by the UK Advisory Committee on Asbestos.

62 NH&MRC 'Report on the Health Hazards of Asbestos', June 1981, Report of Asbestos Ad hoc Subcommittee.

63 ibid.

64 Interview, Julian Peto, Melbourne, 2008. Also C. Rake, C. Gilham, J. Hatch, A. Darnton, J. Hodgson and J. Peto, 'Occupational, domestic and environmental mesothelioma risks in the British population: a case-control study', *British Journal*

of Cancer, 1–9, 2009; Health and Safety Executive, Research Report 646, London, 2009. A 2009 report by the International Agency for Research on Cancer drew the 'fundamental conclusion' that regardless of debate the potency of different asbestos fibres, all forms are carcinogenic. 'Special report: a review of human carcinogens', *The Lancet Oncology*, vol. 10, May 2009.

5   The Baryulgil 'Time Bomb'

1   *Northern Star*, 22 June 1961.

2   On 29 June 1959 the Grafton Branch of the Australian Labor Party asked the Minister for Mines to conduct an inspection at Baryulgil because '... it was reported ... that recently employees at the mine had died from a lung condition and that there was a suspicion that the employees, who are understood to be aborigines [sic] are being exploited.' Trevor Jones, an inspector from the NSW Health Commission reported on 24 May 1960, that although it may be a 'wise precaution' to have Donnelly X-rayed, workers in the plant were exposed to dust of 'low concentrations'. The Baryulgil mine manager reported to Hardie Head Office on 24 March 1960 that the two inspectors '... had lunch here and were altogether very pleasant'.

3   M. Peacock, op. cit., p. 103.

4   JH confidential memo from Dr S. F. McCullagh, 12 January 1971. Hardie later minimised the significance of such high readings because they were short duration 'environmental' samples taken at a fixed point, rather than personal samplers attached to a worker for the duration of a typical shift. But Hardie's medical officer constantly complained about the dust levels at Baryulgil. After a visit in 1974, he described the control of dust as '... far poorer than anywhere else', with asbestos around the bagging operation '... so high as to be readily visible, even with poor illumination'; other Hardie memos described the 'billowing clouds' of dust and 'deplorable' conditions.

5   JH Environmental Control Committee minutes, 19 June 1974.

6   Interview, Tony Mundine, 22 March 2006.

7   Interview, A. Cave, 19 September 2005.

8   E. G. Reeve, chief draftsman, 'Report of Visit to Asbestos Mine, Baryulgil', 28 February 1966.

9   JH Environmental Control Committee minutes, 24 May 1976.

10 T. Hall, *The Ugly Face of Australian Business*, Harper & Row, Sydney, 1980, pp. 34–35.

11 For some of the written history of the area, see George Farwell, *Squatter's Castle: The Story of a Pastoral Dynasty*, Lansdowne Press, Melbourne, 1973; Janet Cannon, *Yugilbar, 1949–1999*, Hardie Grant Publishing, Melbourne, 1999; Jock McCulloch, *Asbestos: Its Human Cost*, University of Queensland Press, Brisbane, 1986.

12 Under government policies, children of 'mixed race' were removed from their families and placed in institutions or church missions. The Baryulgil people still proudly recall the day they defied the welfare officers who had arrived to take their children away.

13 W. Pook, Grafton Base Hospital radiologist, 2 April 1952, nos. 17809–10, reports on chest X-rays of Albert Preece and Harry Mundine.

14 Letter from H. G. Raggat, secretary, Department of National Development, Canberra, to the director, Bureau of Mineral Resources, Melbourne, 27 May 1953, National Archives of Australia, Series A987, Folio 32. See also Series A816.

15 Thyne Reid, chairman's address, extraordinary general meeting, 'James Hardie Asbestos', 24 November 1955.

16 Interview, James Kelso, Sydney, 25 May 2006.

17 Hawke attended the race on 2 October 1983. According to Kelso 'the crowd jumped the fence to talk to him'.

18 See, for example, Stephen Rice, '70 died: claim to asbestos inquiry', *Sydney Morning Herald*, 22 September 1983.

19 Letter from Chris Lawrence, Aboriginal Legal Service, Sydney, to Canadian Asbestos Information Centre, 'Baryulgil Asbestos

Claims', 23 June 1983, quoted in the submission of the South Pacific Asbestos Association to the House of Representatives Standing Committee on Aboriginal Affairs, Inquiry into the Effects of Asbestos Mining on the Baryulgil Community, 1984.

20 Interview, Jock McCulloch, Melbourne, 3 May 2006. McCulloch's book, *Asbestos: Its Human Cost*, University of Queensland Press, Brisbane, 1986, deals extensively with the Baryulgil mine and the inquiry.

21 House of Representatives, 4 October 1983.

22 Interview, Chris Lawrence, 9 February 2007.

23 House of Representatives Standing Committee Inquiry on Aboriginal Affairs, op. cit., transcript of proceedings, 2 December 1983.

24 ibid., 10 February 1984.

25 ibid., 23 August 1984.

26 ibid., 10 February 1984.

27 Dr McCullagh complained in his 1972 annual report that he had long been concerned with asbestos in air levels at Baryulgil and had no doubt that the absence of cases of asbestos disease there was 'a consequence of high labour turnover and not satisfactory workplace hygiene'.

28 Submission of the South Pacific Asbestos Association to the House of Representatives Standing Committee on Aboriginal Affairs, Inquiry into the Effects of Asbestos Mining on the Baryulgil Community, 1984.

29 Interview, James Kelso, op. cit.

30 Deidre Macken, 'Hardie parries health protest', *The Australian*, 28 July 1978.

31 Report by the House of Representatives Standing Committee on Aboriginal Affairs from the Inquiry into the Effects of Asbestos Mining on the Baryulgil Community, 1984, Parliamentary Paper no. 224/1984, AGPS, Canberra, 1984.

32 Samples from the Square taken in early 1989 by Robert Vojakovic, president of the Perth-based Asbestos Diseases Society and analysed

by Dr J. Langley, the scientific officer at the WA Department of Occupational Health, Safety & Welfare, were found to contain 10 per cent white asbestos: Assessment Report, 5 May 1989.

33 Report by the House of Representatives Standing Committee on Aboriginal Affairs, op. cit., ch. 10, par. 10.29.

34 Interview, Lindsay Gordon, 22 March 2006.

35 Human Rights and Equal Opportunity Commission, 'A Report by the Race Relations Commissioner on a Visit to Baryulgil and Malabugilmar', NSW, July 1990.

36 Interview, Dr Ray Jones, Grafton, 20 March 2006.

37 Interview, Robert Vojakovic, Perth, 28 May 2005.

38 G. Hand, 'Background Briefing', ABC National, 20 November 1988.

39 Letter from Gerry Hand, Minister for Aboriginal Affairs, Canberra, to Charles Moran, 31 March 1988.

40 Kelso informed the inquiry that Penge 'is in fact a chrysotile mine', Report by the House of Representatives Standing Committee on Aboriginal Affairs, op. cit., transcript of proceedings, 10 February 1984. p. 1309. In reality, Penge was the world's largest brown asbestos mine.

41 Hardie denied bags from Wittenoom were used at Baryulgil, but Hindle's evidence was corroborated by the former mine manager, Gerry Bourke (notes of interview with G. Bourke by Peter Cashman, 11–12 May 1983, in author's possession); affadavit, William Hindle, Hindle v Marlew Mining Pty Ltd, James Hardie and Coy Pty Ltd, CSR Building Materials Pty Ltd, N8WSC 14382, 1984.

42 Briggs v James Hardie & Coy & Ors, vol. 16 NSWLR 549, 1989. Justice Rogers observed that for the 'undoubtedly afflicted' Briggs '... there seems to me to be something wrong with the state of the law when, in order to recover compensation for his apparent asbestosis, a person in the position of this plaintiff has to mount a challenge to fundamental principles of company law.'

43 Interview, Mark Knight, 13 April 2007.

44 *The Weekend Australian*, 28–29 October 1989, p. 4.

45 'Valuation of asbestos disease related liabilities arising from the
    manufacture or use of asbestos by the former James Hardie entities,
    as at 31 March 2005, KPMG, 14 May 2005, note 10, pp. 85–88.

46 Albert Preece was diagnosed with asbestosis in 1952, less than a
    decade after the mine opened; Simon Tomasic, who worked at
    the mine immediately before its closure in 1979 died from
    another cause, but nonetheless asbestosis was revealed in his
    autopsy.

47 KPMG, op. cit., and updates; DDB communication with author.

48 NSW Health Commission, 'Investigation into the Health of
    the Baryulgil Asbestos Mine Workers', November 1979, 'A
    Re-examination of the Health of the Miners and ex-Miners from
    the Baryulgil Asbestos Mine', 1981, NSW Government Printer,
    Sydney.

49 Personal communication from Ray Jones, Grafton Aboriginal
    Medical Service, 20 March 2006.

50 JH letter, R. Palfreyman to NSW health minister K. Stewart,
    8 February 1978.

51 Report of a meeting at the Bureau of Environmental and Special
    Health Services, NSW Health Commission, held at 9–13 Young
    Street, Sydney, 5 October 1977.

52 Interview, Dr Ray Jones, op. cit.

53 Interview, Brett Freeburn, 30 June 2006.

54 US Institute of Medicine of the National Academies, 'Asbestos:
    Selected Cancers', Committee on Asbestos: Selected Health
    Effects, National Academies Press, Washington DC, 2006.

55 Interview, Dr Anthony Johnston, NSW Dust Diseases Board,
    11 May 2007.

6 Handling the Unions: A 'Delicate Situation'

1 Interview, Bernie Banton, 24 May 2005.

2 Interview, Ray Gietzelt, 29 July 2005. See also his autobiography,
   Ray Gietzelt, *Worth Fighting For*, Federation Press, Sydney, 2004.

3 Gietzelt, ibid.

4 Shanahan's wife, Margaret, told me that a few months before he died, Shanahan left the family home abruptly one afternoon and failed to return. The next day police found him in his car at the edge of the highway, where he had sat all night. He was by then in great pain and may have been contemplating suicide. Gietzelt told me that he organised compensation for Shanahan despite the organiser's treachery.

5 JH memo to executive committee, 'Dust Survey, Camellia, 1957', 24 June 1958.

6 Bob Johnson, 'Doors close on "great place to work",' *Courier-Mail*, 4 December 1992.

7 JH memo to executive committee, 'Dust Survey Camellia, 1957', op. cit.

8 JH letter from Dr McCullagh to the director of NSW Division of Occupational Health, 30 April 1965.

9 JH memo from E. T. Pysden, personnel manager to manager, JH factory at Penrose, New Zealand, 22 June 1966. Once again in this memo Pysden stressed the importance of 'not creating any suggestion of panic ...'

10 KPMG, commissioned by James Hardie, in 2006 estimated that the company would receive 1859 workers' compensation claims, 'Valuation of Asbestos Related Disease Liabilities of Former James Hardie Entities', 2006, p. 106. The true number of those affected would probably be at least twice this, given the propensity of some not to claim and undiagnosed asbestos-associated disease. While the Australian Workers Union covered Hardie employees in WA and Queensland, the Missos had by far the majority of workers, with NSW and Victoria accounting for the most.

11 Interview, Doug Howitt, 22 July 2005. The factory was probably Asbestos Products Pty Limited, a CSR subsidiary which closed in 1957 and later achieved legal prominence in Wren v CSR.

12 *The Miscellaneous Worker*, August 1960.

13 Letter from Doug Howitt, Secretary, NSW Branch, Federated Miscellaneous Workers' Union, to Neil Gilbert, JH Camellia factory manager, 7 December 1965.

14 JH letter from Neil Gilbert, Camellia factory manager, to Doug Howitt, Federated Miscellaneous Workers' Union, 14 February 1966.

15 Memo to state branches from M. O'Brien, general secretary, Australian Railways Union, 11 April 1967. O'Brien circulated the newspaper article '... in case asbestos is in use in railway workshops, and dangers associated with its use has not come to your attention'.

16 *Federation News*, June 1968.

17 *Whyalla News*, 18 November 1968.

18 *Sun*, 4 October 1968.

19 JH memo from Dr S. F. McCullagh to N. Gilbert, 'Biological Effects of Asbestos', 20 November 1968.

20 Interview, Doug Howitt, op. cit. See also letter from N. Gilbert, JH manager, research and development, to D. Howitt, 6 November 1968.

21 Confidential memo from Dr S. F. McCullagh, medical officer, to JH personnel director, 22 August 1972.

22 Interview, Doug Howitt, op. cit.

23 See, for example, the letter from JH director Ray Palfreyman to Ray Gietzelt, general secretary, Federated Miscellaneous Workers' Union, 'Meeting of Experts — The Safe Use of Asbestos', International Labour Organisation, Geneva', 8 February 1974.

24 JH Environmental Control Committee minutes, 6 April 1973. Roberts had previously worked at Wittenoom and would die of lung cancer, but because he had been a smoker his widow went uncompensated.

25 JH memo from N. Gilbert, manager, research and development, to W. Williams, manager, Brooklyn factory, 'Medical Service and Industrial Hygiene', 6 October 1967.

26 Dr S. F. McCullagh, statement, for McKusker v Seltsam Pty Limited and James Hardie & Coy Pty Limited, NSWDDT 179 1996.

27 Interview, Ray Gietzelt, 29 July 2005.

28 NSW Industrial Commission, no. 248, 1975 Asbestos Sheet Makers' (State) Award, before Conciliation Commissioner Manuel, 18 February 1976, pp. 26–27.

29 'Safeguards taken at new MV plant', *Berrimal District Post*, 18 April 1979. See, for example, NSW Division of Occupational Health reports: 'James Hardie Pipe Plant', 30 January 1981; 'James Hardie Pipe Plant, Moss Vale', 19 January 1982.

30 Interview, Ray Gietzelt, op. cit.

31 Interview, Ray Hogan, 15 June 2005.

32 An account of the Melitis case is reported in the book by Michael Cannon, *That Disreputable Firm: The Inside Story of Slater & Gordon*, Melbourne University Press, 1998, pp.115–116.

33 Hardie's Neil Gilbert informed the Brooklyn manager on 13 August 1969 that the company was 'quite happy' for the union to gain access to the records provided it had received clearance from the individual workers, a written promise that had been made by Hardie to the union on 23 May 1966. Yet within a decade Hardie again resisted such access.

34 Interview, Vic Fitzgerald, 3 January 2008.

35 'Asbestos Holocaust', *The Metalworker*, vol. 8, no. 3, April 1987, p. 1. Cook's notes are also still held in the files of the Asbestos Diseases Foundation of Australia at its office at the AMWU headquarters in Granville, NSW.

36 Interview, Armando Gardiman, 28 March 2006.

37 Interview, Ray Giezelt, op. cit.

7   Spinning the Reid 'Legacy'

1   Interview, Ron Hinton, 15 September 2008.

2   J. T. Reid, JHA Annual Report, 1973.

3   J. B. Reid, JHA Annual Report, 1974.

4   J. B. Reid, JHA Annual Report, 1974. Hardie's Super Six asbestos
     fences are still prolific in Perth, where many contain Wittenoom
     blue asbestos that Hardie 'dribbled' through its stock when its
     danger could no longer be concealed.

5   John Reid, JHA Annual Report, 1977.

6   John Reid, JHA AGM, 1978.

7   See, for example, the letter from Alan Bell, director, NSW
     Division of Occupational Health, 'Detection of Cancer of the
     Lung in Asbestos Workers', to James Hardie & Coy Limited,
     14 April 1965.

8   South Australia would later acquire one of the highest per capita
     rates of mesothelioma in the world as a result of asbestos
     exposures during this time.

9   Reid explained in his reports home that the cartel, involving the
     three international asbestos giants, the UK's Turner & Newall,
     Europe's Eternit, and Johns-Manville from the US, was referred to
     by the acronym TEAM. The aristocratic Emsen family directed
     Eternit's role from Belgium and it included six members — Eternit
     Belgian, Italian, Swiss, Austrian, German and French. In Asia,
     TEAM's purpose was to spread the risk in developing countries.
     Each of the three members provided a third of the asbestos to an
     operation to share the financial risk in countries that were not
     'politically or economically stable'. Letter from John Reid, Kuala
     Lumpur, 20 October 1965, Reid Collection, State Library of NSW.

10   ibid.

11   Jock McCulloch, *Asbestos Blues: Labour, Capital, Physicians and
     the State in South Africa*, Indiana University Press, 2002.

12   ibid.; Interview, Richard Meeran, 28 August 2005.

13   Interview, Richard Meeran, ibid.; G. Tweedale and L. Flynn,
     'Piercing the Corporate Veil — Cape Industries and Multinational
     Corporate Liability for a Toxic Hazard 1950–2004', *Enterprise &
     Society*, Oxford Journals, June 2007, vol. 8, pp. 268–296.

14   Nearly 80 per cent of all asbestos mined in the 20th century was
     used after 1960, with 50 per cent used after 1976. Jock

McCulloch and Geoffrey Tweedale, *Defending the Indefensible*, Oxford University Press, 2008, p. 14.

15 Interview, former JH executive who requested anonymity.

16 JH briefing paper prepared for NSW Premier Wran, November 1974. 'Asbestos Cement Pipe Manufacturing Plant — Moss Vale', Reid Collection, State Library of NSW.

17 The Moss Vale factory was sold by Hardie in 1984.

18 Minutes of James Hardie factory managers' conference, 1970, Session Three, 'Today's Marketing Climate — Building Products'.

19 Richard Meeran, 'Cape Plc: South Africa mine workers quest for justice', *International Journal of Occupational Environmental Health*, vol. 9, no. 3, 2003, p. 221.

20 A statement Hardie's director Frank Page adapted from the British Asbestos Information Council and circulated to the media and staff, 24 October 1968, begins: 'Asbestos is by no means alone in having suspicion directed at it for being a possible health hazard. There are hundreds of substances as mundane as fuel oil, charcoal-grilled steak and iron rust which are known or suspected of being cancer causing agents under certain experimental conditions.' Page noted the previous year that Hardie had applied to join the Asbestos Research Council of the UK asbestos manufacturers, who had formed a public relations committee to '… regain a sense of proportion in the public mind' after 'the scare on the ill effects of asbestos has reached almost hysterical proportions', JH memo, 22 August 1967.

21 Ron Bolton, 'Community Relations Proposals', submission to JH board, 1975, Reid Collection, State Library of NSW.

22 Timothy Hall, 'A macabre waiting game', *The Bulletin*, 15 February 1975; JH & Coy memo from Ron Bolton, community relations manager, 14 February 1975.

23 Reid had appointed Heath, who held considerable sway over the young director, but like many in the old guard of Hardie, he was a heavy drinker. He left the company with some bitterness, and his widow later said his death was assocated with an asbestos-

related disease, B. Hills, 'The relentless toll of the widow-maker', *Spectrum (Sydney Morning Herald)*, 3 November 1990. For a full account of David Macfarlane's recollections, see his biography in the National Library, 'From Ichang to Whale Beach', Wendy Willcocks (Books of Writers Network, 2009).

24 This was how Ian Kortlang, former partner of the PR firm Gavin Anderson, described Reid's benevolence. Kortlang later provided PR advice to Meredith Hellicar when she faced charges from the corporate regulator, ASIC. Interview, Ian Kortlang, 2 March 2007.

25 Bill Frew, 'James Hardie Public Relations Action Plan', 1978, Reid Collection, Mitchell Library.

26 Eric White employed Stuart Fist as a computer specialist who in late 1979 searched the world's literature for new research on asbestos dangers, giving Hardie the most sophisticated access to state-of-the-art knowledge. Stuart Fist, interview and correspondence, 17 July 2007.

27 Letter from James Cameron, Neilson McCarthy & Partners to Ron Bolton, JH community relations manager, 17 July 1978, Reid Collection, Mitchell Library.

28 Letter from Bill Frew, Eric White Associates, to Ron Bolton, JH community relations manager, 27 July 1979, Reid Collection, State Library of NSW.

29 When Hardie's Ron Bolton's job as in-house PR was reclassified by JH in 1980 the job description noted: 'An acute awareness is required of the sensitivity of language used in communication both with the community and with employees so as to avoid unfavourable attitudes and possible litigation.' Reid Collection, State Library of NSW.

30 Letter from Bill Frew, Eric White & Associates, to Ron Bolton, JH community relations manager, 29 November 1979, Reid Collection, State Library of NSW.

31 JH & Coy background note, 'Asbestos and Health', January 1979, p. 7. The booklet recommended following the 'simple rules' detailed on Hardie's warning labels — such as minimising dust,

ventilation, use of hand tools, damping down and sealing waste — for this claim to be sustained. The reality was that these procedures were rarely practised on the job, as Hardie's own TV advertisement demonstrated.

32 Letter from Bill Frew, Eric White & Associates to Ron Bolton, JH community relations manager, 10 September 1980, Reid Collection, State Library of NSW.

33 By 1980, Hardie's community relations manager Ron Bolton reported, 'There has been no noticeable decline in sales due to asbestos environmental publicity but nevertheless we believe some sales have been lost (some Government authorities and, architects on specific jobs)', Reid Collection, State Library of NSW.

34 For the official account of the transition from asbestos, see B. Carroll, *A Very Good Business*, James Hardie Industries Limited, Sydney, 1987, pp. 220–221. Other material has been drawn from a variety of interviews with the Hardie staff involved.

35 Maureen Murrill, '$65m. takeover nears an end', Melbourne *Herald*, 13 January 1979.

36 Don Kirkwood, 'Hardie's $100m expansion ends asbestos dependence' *National Times*, 5 January 1980, p. 49.

37 *PM*, ABC Radio, 23 August 1995.

38 Report by JH community relations manager Ron Bolton, 1980, Reid Collections, State Library of NSW.

39 Ron Bolton, 'Corporate Image: The Battle for Hearts and Minds', seminar at Chisholm Institute of Technology, Caufield, 11 May 1982, Reid Collection, State Library of NSW.

40 Report by JH community relations manager Ron Bolton, 1980, Reid Collection, State Library of NSW.

41 Ron Bolton, group community relations corporate advertising project, 1983–1984, Reid Collection, State Library of NSW.

42 JH document, 'Cessation of Asbestos', from information provided by B. Sugg, 21 June 1989, David Jackson, op. cit., exhibit 61.

43 'Goodbye fibro, welcome pino', *Sydney Morning Herald*, 12 November 1983.

44 Ross Greenwood, 'James Hardie and asbestos: a tough bond to break', *Australian Business Review Weekly*, 12–18 November 1984, p. 10.

45 Don Kirkwood, op. cit., p. 49.

46 Letter from Bill Frew to Ron Bolton, JH community relations manager, 27 May 1981, State Library of NSW; Interview, Bill Frew, 6 April 2007.

47 Tony Thomas, 'The high price of inflation', *The Age*, 14 August 1975.

48 ibid.

49 Chanticleer, 'Accountants hear a few words to their discomfort', *Australian Financial Review*, 23 June 1978.

50 Tom Mockridge, 'Reid on shady deal, claim', *Sydney Morning Herald*, 15 October 1982; Bill Hayden, Notice of motion, House of Representatives, 23 September 1982.

51 Don Kirkwood and Nick Yardley, 'The richest men', *Sydney Morning Herald*, 20 July 1980, p. 30.

52 John B. Reid, 'Commonsense Corporate Governance', Australian Institute of Company Directors, Sydney, 2002, p. 88.

53 Robert Pullan, 'Australians "undisciplined" — bi-centennial chief', *Sydney Morning Herald*, 13 May 1979.

54 Young & Rubicam Coudrey, 'A proposal for James Hardie Industries Corporate Advertising', 19 July 1984, Reid Collection, State Library of NSW.

55 Marian Theobald, 'State schools need "private" aid', *Sydney Morning Herald*, 28 August 1985.

56 Deborah Snow and Richard Hubbard, 'Hawke forces Reid to resign', *Australian Financial Review*, 27 September 1985.

57 Obituary, 'David Armstrong, 1941–2006', *Sydney Morning Herald*, 11 January 2007.

58 Ross Greenwood, op. cit., p. 12.

59 Reid proudly recounts his shaping of the Hardie board in his book, John B. Reid, *Commonsense Corporate Governance*, Australian Institute of Company Directors, Sydney, 2002, pp. 77–78.

60 John B. Reid, JH AGM, 1987.

61 Ian Reineke, 'Landmark compensation offer by James Hardie', *Australian Financial Review*, 22 March 1979.

62 Interview, Robert Vojakovic, president Asbestos Diseases Society of Australia, 28 May 2005.

63 Copy in author's possession, 'Discharge', 1982.

64 John B. Reid, *Commonsense Corporate Governance*, op. cit., p. 87.

65 ibid., p. 87.

66 ibid., p. 95.

67 Interview, Armando Gardiman, 28 March 2006.

68 Evidence of Keith Barton, David Jackson, op. cit., 8 June 2004, p. 2698.

69 Confidential letter from J. Balmforth to A. B. Ballment, JH, 24 November 1980, Reid Collection, State Library of NSW.

70 John Balmforth, 'Comments and Information on Centenary Book', Part 5 — 'The John B. Reid Years', 496, 22 July 1987, Reid Collection, State Library of NSW.

71 Confidential letter from J. Balmforth to J. B. Reid, 10 September 1987, Reid Collection, State Library of NSW.

72 Brian Carroll, 'foreword', *A Very Good Business*, James Hardie Industries Limited, Sydney, 1987.

73 File note by Paul Brunton, curator, State Library of NSW, 23 March 1994; Letter from J. B. Reid to Alison Crook, state librarian, State Library of NSW, 15 November 1994.

74 Letter from Keith Barton, managing director, JH, to Paul Brunton, curator, State Library of NSW, 14 July 1999.

75 Cooke v Amaca Pty Ltd and ANOR, 22 October 2008, VSA.

8 A Race Against Time

1 Interview, Allen Drew, 4 December 2008.

2 For a brief history of the Dust Diseases Tribunal written by its first president, see Hon. John O'Meally, 'Asbestos litigation in NSW', *Journal of Law and Policy*, no. 15, 2007, pp. 1209–1223.

3 Interview, Armando Gardiman, 28 March 2006.

4   Dr Maurice Joseph.

5   Interview, DDT official, 2004.

6   Olsen v CSR Pty Ltd, NSWDDT, 72 1994, 16 NSW CCR 56 1998.

7   P. P. McGuinness, 'Judging the Judges', *Sydney Morning Herald*, 13 January 1995.

8   P. P. McGuinness, 'Rights in the wrong hands', *Sydney Morning Herald*, 19 January 1995. This was a bizarre claim that more reflected McGuinness' own ideological bias. It was certainly true that both the federal and WA governments had been keen to see the mine continue production, but while the local Australian Workers' Union also supported its existence, there is no evidence that its view had any impact on CSR's decision to operate. When the company closed Wittenoom the decision was based simply on economic considerations.

9   Many former Hardie executives and advisers, such as Jim Kelso and Mark Knight, confirm this practice was widespread.

10 Interview, David Say, 27 September 2004.

11 Interview, Neil Gilbert, op. cit.

12 Interview, Armando Gardiman, op. cit.

13 ibid.

14 ibid. Later Hardie lawyers would switch tactics, countersuing CSR after an award against it for Belinda Dunne, who had contracted mesothelioma after playing as a child during home renovations involving Hardie asbestos cement containing Wittenoom blue. They argued that CSR should have closed the mine by mid-1965 because of the known health risk. Amaca Pty Ltd v CSR Limited, NSWDDT 18 2008.

15 Interview, Meredith Hellicar, 22 December 2005.

16 Later called Allens Arthur Robinson, but for convenience referred to as Allens.

17 Interview, Armando Gardiman, op. cit.

18 Interview, Mark Knight, op. cit.

19 McCusker v James Hardie Industries, NSWDDT 179, 1996, 1 April 1997, p. 528.

20 Interview, Mark Knight, op. cit.

21 JHA AGM, 1978.

22 Interview, Armando Gardiman, op. cit.

23 The Asbestos Diseases Society of NSW was later to become the Asbestos Diseases Foundation of Australia (ADFA), of which Bernie Banton became the vice-president.

24 Interview, Jeff Shaw, 21 January 2007; Interview, Amando Gardiman, op. cit.

25 Confidential report to the board of James Hardie Industries from Wayne Attrill, litigation manager, and Peter Shafron, general counsel, 26 November 1998. Jackson, op. cit., exhibit 284.

26 ibid.

27 Putt v James Hardie & Coy, James Hardie Industries, NSWDDT 181 1997, CA 40062 1998.

28 The best account of this epic battle has been written by Ben Hills, *Blue Murder*, Sun Books, Sydney, 1989.

29 Wren v CSR, NSWDDT 23 1997, CA NSWLRJ 44 1997, pp. 463–492.

30 Putt v James Hardie and Coy Pty Ltd and ANOR, HC S76 1998. See also Marcus Priest, 'Real sting hidden in twisting tail', *Australian Financial Review*, 1 July 2004.

31 See, for example, the JH board presentation 'Asbestos Trust Structure Discussion', January 2001, which observes that 'notwithstanding corporate veil and Putt ... chance that JHIL may be found liable', Jackson, op. cit., exhibit 121.

32 Wayne Attrill, 'Legal Advice to James Hardie for the Purpose of Actual and Anticipated Litigation — Asbestos Liabilities Management Plan, YEM01 to 03', p. 8. Jackson, op. cit., exhibit 10.

33 Interview, Mark Knight, op. cit.

34 Rolls Royce Industrial Power (Pacific) Ltd v James Hardie & Coy Pty Ltd NSWDDT 5 1999; 18 NSWCCR 653 1999.

35 ibid.

36 In a confidential report to the Board of James Hardie Industries from Wayne Attrill, Litigation Manager and Peter Shafron, general counsel, 2 November 1999, the lawyers observed that, 'We have certainly flushed out the DDT's underlying antipathy to manufacturers, and to JHC in particular', Jackson, op. cit., exhibit 20.

37 Thurbon v James Hardie & Coy Pty Ltd, NSWDDT 7 1999.

38 Interview, Elizabeth Thurbon, 13 November 2003.

39 Confidential report to the board of James Hardie Industries from Wayne Attrill, litigation counsel, 14 September 1999, Jackson, op. cit., exhibit 283.

40 Interview, Elizabeth Thurbon, op. cit.

9   Plotting from the War Room

1 'Project Chelsea, Part B: Communications Strategy in Detail', 2 June 1998, p. 1, briefing paper for James Hardie Industries Board, David Jackson, op. cit., exhibit 61, vol. 3.

2 The family shareholding had already been diluted with the takeover of RCI in 1978. Ron Hinton believed that Reid's divorce settlement a decade later was the reason why it ceased to be the major shareholder. Interview, Ron Hinton, op. cit.

3 Interview, Jim Kelso, 26 May 2006.

4 Lance Norman, 'Legislation call in asbestos case', *Australian Financial Review*, 20 November 1984.

5 Peter Hartcher, 'Asbestosis threat was overstated — Hardie', *Sydney Morning Herald*, 23 November 1984. Kelso also told the newspaper that the asbestos lagging of ship boiler rooms had been very dangerous 'but had not happened in Australia'. In fact, it had. Many Australian ships had asbestos lagging, much of it provided by Hardie, and many Australian navy personnel were to die as a result, including NSW Governor Sir David Martin.

6 A useful insight into the legal tussle with QBE emerged from the litigation briefings prepared for the James Hardie Board tendered in evidence during the Jackson Special Commission, such as the

paper from Shafron and Attrill, 10 May 2000, Jackson, op. cit., exhibit 283.

7 'Standard response to questions about asbestos litigation', board briefing for 1998 AGM, 2 July 1998, p. 10.

8 'Project Chelsea, Part B: Communications Strategy in Detail', op. cit., p. 4, Jackson, op. cit., exhibit 61.

9 M. J. Knight, 'Asbestos Litigation Forecast 1992–2005', James Hardie Industries confidential memo to David Say, 6 April 1992, Jackson, op. cit., exhibit 179.

10 'James Hardie Industries: Review of Potential Liability for Asbestos-related Diseases in Australia as at 31 March 1996', Trowbridge Consulting, 10 October 1996, Jackson, op. cit., exhibit 2, vol. 3, tab 12, p. 589.

11 Email from P. Shafron, JH general counsel to P. Cameron, J. Martin, Allens, 'Trowbridge', 3 June 2000, Jackson, op. cit., exhibit, 42.

12 Steve Ashe, JH vice-president, pubic affairs, 'Summary of Analysts' Valuation of Asbestos Liability', David Jackson, op. cit., exhibit 121. Hardie's general counsel candidly observed to the board that the level of the company's asbestos liabilities was 'quite possibly worse than it has been leading the market to believe', P. Shafron, 'Asbestos Board Paper', 4 February 2000, p. 9, Jackson, op. cit., exhibit 22.

13 'Project Chelsea — Board Subcommittee meeting', 31 March 1998, 2.3 'Asbestos', SBC Warburg Dillon Read meeting notes, Jackson, op. cit., exhibit 61.

14 Confidential JH memo from Peter Shafron to Peter Macdonald, Phillip Morley and Greg Baxter, 1 February 2001, Jackson, op. cit., exhibit 121.

15 Letter from Phillip Morley, JHI chief financial officer, to US SEC, 14 April 1988, p. 5, Jackson, op. cit., exhibit 91.

16 'Project Chelsea, Part B: Communications Strategy in Detail', op. cit.

17 See Chapter 7.

18 Paul Cleary, '$1 bn revenue lost to tax havens', *Sydney Morning Herald*, 27 April 1989.

19 Elizabeth Knight, 'Strong hands seek to guide the fortunes of the good ship Hardie', *The Australian*, 18 September 1993.

20 'Project Chelsea, Part B: Communications Strategy in Detail' op. cit.

21 ibid.

22 Summarised in a letter from PricewaterhouseCoopers to Allens, 5 November 1999, Jackson, op. cit. exhibit 61, vol. 4, tab 5, pp. 19–25.

23 Draft confidential JH memo from Peter Shafron to Peter Macdonald, 'Asbestos Liability — Continuous Disclosure', 11 October 2000, Jackson, op. cit., exhibit 150.

24 Hardie had conflicting advice on its disclosure requirements under the new standard, with Allens insisting that future unknown asbestos claims still need not be mentioned. The company knew that its liabilities could be '… quite possibly worse than it has been leading the market to believe' (Peter Shafron, JHIL board paper 'Asbestos', 17 February 2000, Jackson, op. cit., exhibit 283, vol. 5.) But CSR's plan to disclose its estimated future liability greatly increased pressure on Hardie to do likewise.

25 Peter Cameron, 'James Hardie Industries Limited — Ultimate Resolution', 15 April 1999, Jackson, op. cit., exhibit 121.

26 Peter Shafron, 'Big Picture Options for James Hardie's Asbestos Liabilities in Australia', 19 November 1999, Jackson, op. cit., exhibit 61, vol. 4, tab 9.

27 Interview, Meredith Hellicar, 22 December 2005.

28 Peter Shafron, op. cit.

29 Phillip Morley, JHI board paper, 'Potential Separation Structure Outline', 9 December 1999, Jackson, op. cit., exhibit 21.

30 'If the funds prove to be inadequate, will JHINV [the Netherlands parent company] put in more money?' Gill asked. 'No,' Shafron replied. Notes of the teleconference at Phillips Fox, Michael Gill, Peter Shafron, Steve Ashe and others, 18 July 2000, Jackson, op. cit., exhibit 61.

31 Confidential memo from Michael Gill to James Hardie, 4 August 2000, Jackson, op. cit., exhibit 61, p. 153.

32 JHI email from Peter Shafron to David Minty, Trowbridge, 14 April 2000, Jackson, op. cit., exhibit 50, no. 6.

33 Email from Trowbridge's David Minty to Peter Shafron, 13 April 2000, Jackson, op. cit., exhibit 50, no. 6.

34 JHI email from Wayne Attrill to Peter Shafron, 20 April 2000, Jackson, op. cit, exhibit 57, vol. 1, p. 109.

35 Peter Shafron repeatedly urged staff, directors and consultants to shred or delete any non-current documents on this and later projects to separate the company's asbestos liabilities, reflecting an early warning from Hardie's legal advisers that plaintiff lawyers might one day subpoena records. See JHI board paper 'Confidentiality', 7 June 2000, Jackson, op. cit., exhibit 283; Peter Shafron email to Peter Cameron and David Robb, Allens, 21 March 2001, 'Please do not distribute further nor retain copies once read', Jackson, op. cit., exhibit 194. JH email from Shafron to group management team, 'Confidentiality-/Project Green', 9 June 2000, Jackson, op. cit., exhibit 283.

36 JHI email from Peter Shafron to Wayne Attrill, 16 April 2000, Jackson, op. cit., exhibit 57, vol. 1, p. 150; W. Attrill, evidence, David Jackson, op. cit., 29 April 2004.

37 According to Commissioner Jackson, Shafron 'sought to exercise very great influence' over the Trowbridge report, Jackson, op. cit., findings, p. 212. See also exhibit 57, vol. 1, pp. 144, 229–255.

38 Peter Shafron, evidence, 12 May 2004, Jackson, op. cit., p. 1586.

39 Confidential fax from Roy Williams, Allens, to Wayne Attrill, James Hardie, 23 June 2000. Jackson, op. cit., exhibit 61, vol. 9, tab 13.

40 W. Attrill, evidence, Jackson, op. cit., 4 May 2004, p. 1192.

41 P. Cameron, Allens, 'Project Green, Advice on Structure and Separation Issues, Part 2, Separation: Asbestos Risks and Liability, Par 3, Buffer Issues', 5 April 2000, Jackson, op. cit., exhibit 283.

42 Steve Ashe, 'Review of the Draft Trowbridge Report in the Context of Stakeholder Management', 8 August 2000, Jackson, op. cit., exhibit 61, p. 158.

43 Steve Ashe, 'Project Green: Stakeholder Management', Jackson, op. cit., exhibit 61, p. 185.

44 Notes by Wayne Attrill, meeting with Tillinghast, 23 August 2003, Jackson, op. cit., exhibit 61, vol. 4, tab 50, pp. 324–326.

45 Bruce Watson and Mark Hurst, 'Asbestos Liabilities', Deloitte Touche Tohmatsu, VIII Accident Compensation Seminar, 29 November 2000.

46 JH email from Peter Shafron to Peter Macdonald, 'November Asbestos Developments', 1 December 2000, Jackson, op. cit., exhibit 57, vol. 4, p. 795.

47 ibid.; email from Peter Shafron to Wayne Attrill, 2 December 2000.

48 Confidential JH memo from Wayne Attrill to Peter Shafron, 4 December 2000, Jackson, op. cit., exhibit 150, p. 21.

49 Greg Baxter, draft news release, 18 December 2000, Jackson, op. cit., exhibit 61, p. 49.

50 Confidential JH email from Peter Shafron to Peter Macdonald, 'Green: Stakeholder issues', 5 October 2000, Jackson, op. cit., exhibit 150.

51 JH email from Steve Harman to Peter Shafron, 7 October 2000, Jackson, op. cit., exhibit 150.

52 Confidential JH email from Peter Shafron to David Robb, Allens, 20 September 2000, Jackson, op. cit., exhibit 146.

53 Confidential JH email from Stephen Harman to Peter Macdonald, Peter Shafron, Phillip Morley, Greg Baxter and Steve Ashe, 18 February 2001, Jackson, op. cit, exhibit 72.

54 Interview, Dennis Cooper, 23 January 2008.

55 JH email from Peter Shafron to Peter Macdonald, 10 January 2001, Jackson, op. cit., exhibit 150; JH conference call, 4 January 2001, Jackson, op. cit., exhibit 203.

56 JHI facsimile from Peter Macdonald to Sir Llew Edwards, 3 January 2001, Jackson, op. cit., exhibit 121.

57 Handwritten notes of JHI conference, Jackson, op. cit., exhibit 203.

10 The 'Fully Funded' Foundation

1 'NSW Government Stakeholder Strategy, Current as of January 11 2001', transcript of workshop session between Hawker Britton's Lisa-Jayne Loch and Greg Baxter of JHI, copy in author's possession.

2 Putt v James Hardie and Coy Pty Limited and ANOR, HC S76 1998.

3 Wayne Attrill, 'Legal Advice to James Hardie for the Purpose of Actual and Anticpated Litigation — Asbestos Liabilities Management Plan, YEM01 to 03', p. 13, Jackson, op. cit., exhibit 10.

4 Interview, David Britton, 23 January, 2008.

5 'Communications Strategy', confidential draft, email from Greg Baxter to Peter Macdonald, Peter Shafron, 'LJ' Loch, Jane Rotsey, 3 February 2001, p. 1, Jackson, op. cit., exhibit 283.

6 Confidential JHI email from Peter Shafron to Peter Macdonald, 1 February 2001, Jackson, op. cit., exhibit 189.

7 Confidential JHI email from Peter Shafron to Roy Williams, Allens, 25 January 2001, Jackson, op. cit., exhibit 75, vol. 7, tab 100.

8 Evidence from JHI's Peter Shafron and Trowbridge's David Minty and Keith Marshall, Jackson, op. cit.

9 Evidence of Wayne Attrill Jackson, op. cit., 5 May 2004, pp. 1225–1226.

10 'James Hardie Industries Limited — Asbestos Litigation', Trowbridge Consulting, 13 February 2001, Jackson, op. cit., exhibit 50, tab 18, p. 160.

11 Report by Greg Baxter, Jackson, op. cit., exhibit 189, vol. 1, p. 187.

12 So described by Commissioner David Jackson, Jackson, op. cit., report, vol. 1, p. 12.

13 JHI board meeting, 17 January 2001, Jackson, op. cit., exhibit 187, vol. 1, tab 9; Robb, evidence, p. 2830.

14 JHI email from Peter Shafron to David Robb, Allens, 9 February 2001, Jackson, op. cit., exhibit 215.

15 David Minty, supplementary statement, 21 April 2004, Jackson, op. cit., exhibit 15.

16 David Robb and Peter Cameron, evidence, Jackson, op. cit.

17 Interview, Meredith Hellicar, op. cit.

18 Interview, Dennis Cooper, op. cit.

19 ibid.

20 Greg Baxter, 'Communications Strategy', Project Green board paper, Jackson, op. cit., exhibit 42.

21 ibid.

22 ibid.

23 Interview, David Britton, 23 January 2008; also JH notes from teleconference 6 January 2001, Jackson op. cit, exhibit 144.

24 Formerly called Gavin Anderson Kortlang.

25 Loosley refused to confirm this, claiming 'commercial-in-confidence', but others inside James Hardie or employed as consultants at the time assert that he continued to advise the company during the Jackson Special Commission of Inquiry. Grey says he never accepted the consultancy, although Hardie board reports claim he had. He says he simply told Hardie to fund more research into asbestos diseases.

26 Evidence of Phillip Morley, Jackson, op. cit., 1 June 2004, pp. 2252–2253.

27 Letter to JHI from Ewen Waterman, Access Economics, 14 February 2001; Letter to JHI from David Brett, PricewaterhouseCoopers, 14 February 2001, Jackson, op. cit., exhibit 1.

28 JH email from Peter Macdonald to Greg Baxter, 10 February 2001. Jackson, op. cit., exhibit 145.

29 JH board presentation, February 2002, p. 28, Jackson, op. cit., exhibit 42. AJO 121. See also Allen's notes from a briefing on the meeting, Jackson, op. cit., exhibit 136.

30 Greg Baxter, 'Communications Strategy', Project Green board paper, Jackson, op. cit., exhibit 283.

31 JHI media release, 'James Hardie Resolves its Asbestos Liability Favourably for Claimants and Shareholders', 16 February 2001.

32 Commissioner David Jackson, QC, Report of the Special Commission of Inquiry into the Medical Research and Compensation Foundation, vol. 1, p. 358.

33 Anthony Hughes, 'Hardie Cleans Up Its Asbestos Act', *Sydney Morning Herald*, 17 February 2001.

34 Ben Hills and Anthony Hughes, 'Asbestos money "not enough"', *Sydney Morning Herald*, 17 February 2001, p. 2.

35 Ben Hills, 'Every breath you take', *Sydney Morning Herald*, 27 February 2001, p. 11.

36 Letter from Peter Macdonald, JHI CEO, to Ben Hills, *Sydney Morning Herald*, 22 February 2001, Jackson, op. cit., exhibit 61.

37 'We are expecting a very adverse article this Saturday on the asbestos separation by a Mr Ben Hills,' emailed Shafron to Trowbridge's Minty and Marshall. 'Our preferred position is that no one from Trowbridge talks to Mr Hills ...', JHI email from Peter Shafron, 22 February 2001, Jackson, op. cit., exhibit 55.

38 Email from Roy Williams, Allens to Peter Shafron, JHI, 28 February 2001, Jackson, op. cit., exhibit 163.

39 Email from Peter Shafron, JHI, to Roy Williams, Allens, 1 March 2001, Jackson, op. cit., exhibit 163.

40 Evidence of Wayne Attrill, Jackson, op. cit., 3 May 2004, p. 1094.

41 Interview, Greg Baxter, 7 March 2006.

42 Media release from the minister for industrial relations, John Della Bosca, 'Minister Welcomes Asbestos Fund', 16 February 2001.

43 JH email from Peter Macdonald, 18 February 2001, Jackson, op. cit., exhibit 74.

44 Letters from Paul Bastian, AMWU, Bernie Riordan, CEPU, Barry Robson, MUA, to Bob Carr, premier, Bob Debus, attorney-general and John Della Bosca, minister for industrial relations, 21 February 2001, AMWU files, Granville.

45 Interview, Bob Carr, 12 March 2008.

46 See, for example, AAP story, 'Unions plea for help in compo fight for asbestos victims', 22 February 2001; 'Asbestos payout plea', *Daily Telegraph*, 23 February 2001; ABC Radio News, Sydney 702, 7.45am, 7 March 2001.

47 JHI email from Steve Ashe to Stephen Loosley, 28 February 2001, Jackson, op. cit., exhibit 135.

48 JHI letter from Steve Ashe to Jan Primrose, AMWU, 9 March 2001. Jackson, op. cit., exhibit 135.

49 Greg Baxter, 'Summary of meeting with NSW Dust Diseases Board — Thursday 15 March 2001, Jackson, op. cit., exhibit 135.

50 Interview, Greg Combet, 4 January 2008.

51 Interview, Bill Mansfield, 4 January 2008.

52 Greg Baxter, 'Summary of Meeting with ACT — Friday, 16 March 2001, Jackson, op. cit., exhibit 187.

53 Interview, Greg Combet, op. cit.

54 JH media release, 'Foundation Safeguards Rights', email from Greg Baxter to Stephen Loosley, 21 March 2001, Jackson, op. cit., exhibit 135.

55 Interview, Dennis Cooper, op. cit.

56 ibid.

57 JH memo from Peter Shafron to Don Cameron, 20 March 2001, Jackson, op. cit., exhibit 42.

58 Confidential JHI email from Peter Macdonald to Peter Shafron, copied to Greg Baxter and Phillip Morley, 23 April 2001. Jackson, op. cit., exhibit 150.

59 Evidence of Sir Llew Edwards, Jackson, op. cit., exhibit 13, p. 35.

60 Interview, Dennis Cooper, op. cit.

61 Confidential JHI memo from Peter Shafron to Peter Macdonald, 'Draft: JHIL Post Green', 23 March 2001, p. 2, Jackson, op. cit., exhibit 147.

62 ibid., p. 3.

63 ibid.

64 Robin Bromby, 'Hardie goes Dutch', *The Australian*, 25 July 2001.

65 Justice Santow, NSWSC 741, 28 August 2001, Jackson, op. cit., exhibit 224, vol. 2, tab 37, p. 480.

66 Letter from Peter Cameron and David Robb, Allens, to Dorothy Laidler, Associate to Mr Justice Santow, NSW Supreme Court, 13 August 2001, p. 9, Jackson, op. cit., exhibit 61.

67 Evidence of David Robb, Jackson, op. cit., 10 June 2004, p. 2885.

68 JH email from Peter Shafron to David Robb, 27 March 2001, Jackson, op. cit., exhibit 194.

69 Justice Santow, Jackson, op. cit., exhibit 224, vol. 2, tab 37, p. 481.

70 Letter David Robb, Allens, to Dorothy Laidler, associate to Mr Justice Santow, NSW Supreme Court, Jackson, op. cit., exhibit 61, vol. 6, tab 24, p. 3.

71 Justice Santow, Jackson, op. cit., exhibit 224, vol. 2, tab 37, p. 481.

72 ibid.

73 ibid., and evidence of David Robb, 11 June 2004, Jackson, op. cit., p. 2885.

74 Trowbridge, report of MRCF asbestos liabilities, August 2001, Jackson, op. cit., exhibit 2, vol. 3, tab 8. Steve Ashe reported to his colleagues that Dennis Cooper told him 'the directors are all walking around with very long faces'. Confidential JH email from Ashe to Peter Shafron, 7 August 2001, Jackson, op. cit., exhibit 189.

75 Letter from Sir Llew Edwards, MRCF, to Peter Macdonald, JHI, 24 September 2001, Jackson, op. cit., exhibit 121, vol. 8, no. 8.

76 Peter Macdonald, evidence, Jackson, op. cit., p. 2396. Commissioner Jackson rejected Macdonald's claim not to have read the letter before 24 October, Jackson, report, vol. 1, p. 394.

77 JHIL board meeting, Regent Hotel, Sydney, 28 September 2001. Jackson, op. cit., exhibit 283.

78 Notes of David Robb, Allens, 30 January 2002, Jackson, op. cit., exhibit 187.

79   Evidence of Alan McGregor, 11 May 2004, Jackson, op. cit., p. 1506.

80   Notes by the MRCF's Dennis Cooper of the meeting, Jackson, op. cit., exhibit 7, MRCF 2, tab 11, pp. 16B–16C.

81   'Minutes of meeting of representatives of Amaca Pty Limited [MRCF] and James Hardie', 22 March 2002, Qantas Club Lounge, Sydney Domestic Airport, Jackson, op. cit., exhibit 189.

82   File notes by Allens' lawyer Julian Blanchard, 25 March 2002, Jackson, op. cit., exhibit 302.

83   Jenni Priestley, Allens file note, meeting with JHI's Phillip Morley, Allens' David Robb and Julian Blanchard and others, 12 April 2002, Jackson, op. cit., exhibit 187.

84   Blanchard file notes, meeting between Allens lawyers Julian Blanchard, David Robb and Michael Ball, 17 July 2002, Jackson, op. cit., exhibit 302.

85   ibid.

86   ibid.

87   Blanchard file notes, telephone conference between JHI's Peter Shafron and Allens lawyers David Robb and Julian Blanchard Jackson, op. cit., exhibit 302.

88   Peter Shafron, confidential JHI board paper, 'ABN transfer and cancellation of partly paid shares', 3 February 2003, Jackson, op. cit., exhibit 148, vol. 2, tab 24, p. 531.

89   Notes by Dennis Cooper, of MRCF meeting with JHI's Peter Macdonald, Peter Shafron and Phillip Morley, 16 July 2002, Jackson, op. cit., exhibit 150.

90   Interview, Dennis Cooper, op. cit.

91   File note by Dennis Cooper, MRCF, 21 January 2003, Jackson, op. cit., exhibit 7.

92   Damon Kitney, 'Hardie to hand back extra cash', *Australian Financial Review*, 3 February 2003.

93   Confidential JHI email from Peter Macdonald to Peter Shafron, Greg Baxter and others, 18 March 2003, Jackson, op. cit., exhibit 150.

94 Peter Macdonald, 'Draft Analysis — Options to Manage
ABN 60 ...', 17 March 2003, Jackson, op. cit., exhibit 61.

95 JHI email from Peter Macdonald to Peter Shafron, Greg Baxter,
Phillip Morley and David Robb, 26 March 2003, Jackson,
op. cit., exhibit 43.

11 Behind the Scenes: Asbestos Politics

1 Edwards told JH's Peter Macdonald at a meeting on 16
December 2002 that 'Reform was appearing more likely but
was not possible till after the NSW State election', notes of
Dennis Cooper, Jackson, op. cit., exhibit 7; also in JHI email
from Peter Macdonald to Peter Shafron, 23 January 2003: 'Sir
Llew also commented that he had continued to work politically
to see if improvement can be achieved to the claims process
with the DDT and plaintiffs bar. Sir Llew is very hopeful that
significant improvement to current costs will be achieved later
this year following the NSW state elections', Jackson, op. cit.,
exhibit 75.

2 Interview, Ian Hutchinson, 17 May 2007.

3 ibid.

4 BIL's director on the JHIL board even asked when the idea of the
foundation was being being canvassed whether one of its duties
could be 'to seek a no-fault compensation regime from
government', JH email from Peter Shafron to Allens lawyers Peter
Cameron and David Robb, 'Dan O'Brien issues', 10 January
2001, Jackson, op. cit., exhibit 89.

5 See, for example, Edwards letter to Macdonald on 24 September
2001, '... it is our strong resolve to implement a political strategy
aimed at refining and possibly changing the compensation
system'. Jackson, op. cit., exhibit 7.

6 Interview, Sir Llew Edwards, 22 July 2004.

7 ibid., the reference to Labor 'mates' presumably meant Frank
Walker, a former minister in both NSW and federal Labor
governments, who was later appointed a judge at the DDT.

8   The detailed letter, in the author's possession, was believed to have been drafted by a senior officer of Allianz on behalf of the Insurance Council of Australia, although the signature is missing from the leaked copy.

9   Interview, Bob Carr, op. cit.

10   Report by Peter Macdonald to JHI board, 13 February 2003, Jackson, op. cit., exhibit 153.

11   Rolls-Royce Industrial Power (Pacific) Ltd v James Hardie & Coy Pty Ltd, NSWDDT 5 1999; 18 NSW CCR 653 1999.

12   Report by Peter Macdonald to JHI board, 14 May 2002, Jackson, op. cit., exhibit 153.

13   'A Proposal for Stakeholder Management Assistance', letter from David Minty, Trowbridge, to MRCF, 26 March 2003.

14   Interview, Roger Wilkins, 28 February 2008.

15   ABC Radio News, 7 June 2003, 2.00 pm.

16   Allianz Australia media statement, 11 June 2003.

17   AAP bulletin, 'NSW Dust Disease Tribunal slammed in Qld', 18 June 2003.

18   Linda Morris, 'Groups cough at asbestos claims shake-up', *Sydney Morning Herald*, 14 July 2003.

19   Interview, Bob Debus, 24 March 2008.

20   ibid.

21   Interview, Nancy Milne, 17 May 2007.

22   Interview, John Della Bosca, 10 February 2006.

23   Peter Macdonald, 'Governance should be seen as a chance to restore credibility', *Australian Financial Review*, 14 September 2003.

24   In 2003 Baxter and JHI won *IR* magazine's Investor Relations Awards for 'Overall Investor Relations', 'Best Investor Relations Officer', 'Best Communication of Shareholder Value', 'Best Ongoing Management of Continuous Disclosure', 'Best Investment Community Meetings', 'Best Corporate Literature', 'Best Use of the Internet for Investor Relations', as well as rating first in Australia on the Ross Carmichael Singer 'Corporate

Confidence Index' for Overall Investor Communication,
Information Disclosure, Information Analysis and Briefing, and
Investor Relations website.

25 Reid spoke at the 2003 annual conference of the Australian
Institute of Company Directors, 14–16 May, as a panelist
debating the insurance and litigation 'crisis'.

26 John B. Reid, *Commonsense Corporate Governance*, Australian
Institute of Company Directors, Sydney, 2002.

27 MRCF file note, 'Meeting between foundation and James Hardie
held on Thursday, 9 October 2003, at the Westin Hotel, Jackson,
op. cit., exhibit 7, tab 11.

28 ibid.

29 MRCF, media release, 'Asbestos Compensation Funding',
29 October 2003, Jackson, op. cit., exhibit 154.

30 Peter Macdonald, JH teleconference, 30 October 2003.

31 John Della Bosca, NSW Minister for Industrial Relations, media
release, 'Medical Research and Compensation Foundation" 30
October 2003.

32 Interview, John Della Bosca, op. cit.

33 Interview, Roger Wilkins, 3 March 2008. Wilkins was not present
at the visit, and an impassioned plea for an investigation with
Royal Commission powers was made by Michael Slattery, the
foundation's QC, which appeared to sway the Cabinet Office
deputy director-general, Leigh Sanderson. Interviews, Nancy
Milne and Ian Hutchinson, 17 May 2007.

34 Greg Baxter, 'Communications Strategy', Project Green board
paper, Jackson, op. cit., exhibit 80.

35 Interview, Peter Macdonald, 13 November 2003.

36 Interview, Meredith Hellicar, op. cit.

37 Interview, Bob Debus, op. cit.

38 Interview, Meredith Hellicar, op. cit.

39 Interview, John Della Bosca, op. cit.

40 Peter Macdonald, *PM*, ABC Radio, 16 March 2004.

41 Jörg Probst, statement, Jackson, op. cit., exhibit 231.

42 Interview, Nancy Milne, op. cit.
43 JH email from Peter Macdonald to Peter Shafron and David Robb, 1 August 2001, Jackson, op. cit., exhibit 152. Also Elisabeth Sexton, 'Hardie manoeuvre worried CSR', *Sydney Morning Herald*, 4 June 2006.
44 Interview, Nancy Milne, op. cit.
45 ibid.
46 Interview, Bernie Banton, 25 May 2005.
47 Michael Gill, supplementary statement, 21 May 2004, Jackson, op. cit., p. 9.
48 Interview, Nancy Milne, op. cit.
49 ibid.
50 Interview, Elizabeth Thurbon, 11 March 2008.
51 Peter Shafron, evidence, 7 May 2004, Jackson, op. cit., p. 1370.
52 ibid., 13 May 2004, pp. 1728–1729.
53 'What a Client', *Sydney Morning Herald*, 4 June 2004.
54 Peter Macdonald, evidence, 7 June 2004, Jackson, op. cit., p. 2628.
55 Interview, Peter Gordon, 12 April 2007.
56 Interview, Greg Combet, op. cit.
57 Interview, Greg Baxter, op. cit.
58 ibid.
59 Commissioner David Jackson, 7 July 2004, Jackson, op. cit.
60 David Robb, evidence, 10 June 2004, Jackson, op. cit., p. 2873.
61 Interview, Tanya Segelov, op. cit.
62 Interview, Bernie Banton, op. cit.
63 KPMG, 'James Hardie Actuarial Expert Witness Report', 4 June 2004.
64 Elisabeth Sexton, 'Hardie warns on law suits', *Sydney Morning Herald*, 15 July 2004.
65 Edwards & Ors v Attorney-General & ANOR, NSW CA 272 revised – 6/08/2004.
66 Interview, Greg Combet, op. cit.
67 Ean Higgins and Paddy Manning, 'Fund a bid to "blackmail the dying"', *The Australian*, 29 July 2004, p. 6.

68 Greg Combet, 'Working together in the interests of the country', Sydney Institute, 5 August 2004.

69 Interview, Greg Combet, op. cit.

70 Anthony Meagher, QC, James Hardie counsel, Jackson, op. cit., pp. 4041–4046.

71 Interview, Meredith Hellicar, op. cit.

72 Meredith Hellicar, interview with Stephen Long, *AM*, ABC Radio, 17 August 2004.

73 Interview, Meredith Hellicar, op. cit.

74 Peter Gosnell, 'Victims' spokesman reckons James Hardie simply moving the deckchairs on the Titanic', *Daily Telegraph*, 19 August 2004.

75 Ean Higgins, 'Carr may support Hardie workers', *The Australian*, 18 August 2004.

76 Interview, Bernie Banton, op. cit.

77 Meredith Hellicar, address to James Hardie Industries shareholders, 15 September 2004.

78 Interview, Armando Gardiman, op. cit.

79 Interview, Bob Carr, op. cit.

80 Interview, Bernie Banton, op. cit.

81 Interview, Bob Debus, op. cit.

82 Interview, Greg Combet, op. cit. Wilkins later told me he had no recollection of the incident.

83 Interview, Meredith Hellicar, op. cit.

84 Interview, Roger Wilkins, 22 February 2008.

85 See, for example, the minutes of JH's Building Products Committee, June and July 1981; Wunderlich, letter from Queensland state manager to managing director, 7 August 1970, 'Asbestos Rule — Factories and Shops Act'.

86 Interview, Wilkins, op. cit.

87 Marcus Priest, 'Carr feels the heat on Hardie', *Australian Financial Review*, 20 September 2004.

88 Interview, Roger Wilkins, op. cit.

89 Interview, Bob Carr, op. cit.

12 Striking a Deal

1   David Jackson, op. cit., pp. 223, 359, 367, 387, 419, 546.

2   Interview, Bernie Banton, 24 May 2008.

3   Jackson, op. cit., p. 456.

4   ibid., p. 367.

5   ibid., p. 517.

6   ibid., p. 567–568.

7   Interview, Bob Carr, op. cit.

8   Interview, Greg Combet, op. cit.

9   Interview, Meredith Hellicar, op. cit.

10 Interview, Greg Combet, op. cit.

11 Interview, Armando Gardiman, 10 January 2008.

12 Interview, Bernie Banton, op. cit.

13 ibid.

14 Interview, Greg Combet, op. cit.

15 ibid.

16 Meredith Hellicar, JH media conference, 25 October 2004.

17 Interview, Sir Llew Edwards, 25 November 2004.

18 Like a number of ongoing and former James Hardie employees or consultants, this source was bound by a confidentiality agreement with significant penalties if breached, and for that reason requested his identity be withheld.

19 Interview, Greg Combet, op. cit.

20 ibid.

21 Bob Carr, NSW premier, news release, 6 October 2004.

22 Interview, Karen Banton, 14 December 2006.

23 Bob Carr, NSW premier, news release, 'Asbestos Legal Costs review'; JH media release, 'James Hardie welcomes asbestos legal costs review', 18 November 2004.

24 Medical Research and Compensation Foundation, media release, 25 November 2004.

25 Bob Carr, NSW premier, media conference, 25 November 2004.

26 Personal communication, John Wells.

27 Meredith Hellicar, ABC Radio, 24 November 2004.

28 Meredith Hellicar, interviewed by Alan Jones, 2GB Radio, 26 November 2004.

29 Interview, Bob Carr, op. cit.

30 Interview, Nancy Milne, op. cit.

31 Alan Jones, 2GB Radio, 8 December 2004.

32 Interview, Bernie Banton, op. cit.

33 Bob Carr, NSW premier, news release, 7 December 2004.

34 Interview, Bernie Banton, 21 December 2004.

35 Dust Diseases Tribunal (Standard Presumptions — Apportionment Order 2007), NSW Government, 23 March 2007.

36 Elisabeth Sexton, 'Premier puts heat on Hardie', *Sydney Morning Herald*, 15 August 2005.

37 ABC News Online, 'Union pressures James Hardie to asbestos compo', 18 August 2005.

38 Roz Alderton, 'Hardie to cut a deal — $4.5 bn asbestos payout in limbo', *Daily Telegraph*, 20 August 2005.

39 This source, still in the NSW public service, prefers to remain anonymous.

40 Elisabeth Sexton, 'Threat to enforce asbestos payout', *Sydney Morning Herald*, 21 November 2005.

41 Interview, John Della Bosca, op. cit.

42 Interview, Greg Combet, op. cit.

43 Marcus Priest, 'Profit will end the pain', *Australian Financial Review*, 2 December 2005.

44 Letter from R. Chenu, CFO, JH to Michael Gill, publisher, *Australian Financial Review*, 2 December 2005; Correction, *Australian Financial Review*, 5 December 2005; Interview, Michael Gill, 31 March 2008.

45 James Hardie, media release, 'ATO decision on tax exempt status of SPF', 23 June 2006.

46 Paul Mulvey, 'Hardie campaign takes its toll on an exhausted Banton', AAP, 4 August 2006.

47 Heath Aston, 'Block Hardie increase', *Daily Telegraph*, 25 September 2006.

48 'James Hardie directors seek pay rise', AAP, 19 September 2006.

49 Letter from Bob Debus, NSW attorney-general to Sir Llew Edwards, chairman, Medical Research and Compensation Foundation, 12 January 2005. Annexure to affidavit by Ian Hutchinson, 31 January 2005, Edwards & Ors v NSW attorney-general, NSW SC 3608 2004.

50 Interview, Ian Hutchinson, op. cit.

51 Letter from Bob Debus, NSW attorney-general to Sir Llew Edwards, chairman, Medical Research and Compensation Foundation, 27 February 2006.

52 The new fees provided the three directors with a total of $328,000 pa, compared with a total of $786,600 for the previous four directors, plus travelling expenses for Sir Llew Edwards, confidential letter from Ian Hutchinson, chairman, Medical Research and Compensation Foundation, to Leigh Sanderson, deputy director-general, NSW Cabinet Office.

53 Interview, Greg Combet, op. cit.

54 Interview, Nancy Milne, op. cit.

55 'No Glory for James Hardie', editorial, *Sydney Morning Herald*, 9 February 2007.

56 James Sheller, 'Claims Resolution Process — Areas requiring resolution', Dust Disease conference, UNSW Law School Centre of Continuing Education, 26 February 2009.

57 ibid.

58 Armando Gardiman, 'Dust Diseases Litigation in NSW, The Claims Resolution Process — Time for Reform', NSW State Legal Conference, 22 August 2008; Interview, Gardiman, 15 September 2008.

59 Judge W. P. Kearns, SC, 'Asbestos Litigation in NSW', NT Law Society, August 2008, p. 7; Stewart v QBE Insurance (Aust.) Ltd and Wallaby Grip Limited, NSWDDT 7279 2007, 4 December 2007; Dawson v Amaca Pty Limited, NSWDDT 7152 2007, 29 January 2008; Evans v Amaca Pty Limited and Seltsam Pty Limited and QBE Insurance (Aust.) Ltd, NSWDDT 7346 2007, 3 March 2008.

13 Under the Carpet: Unfinished Business

1   Interview, Meredith Hellicar, op. cit.

2   Interview, Neil Gilbert, op. cit.

3   Wayne Attrill, 'Asbestos Liabilities Management Plan YEM01 to 03', Jackson, op. cit., p. 14, exhibit 10.

4   JH conference between Peter Shafron, Wayne Attrill, Michael Gill and Russell Adams, 25 July 2000, Jackson, op. cit., exhibit 100.

5   JH report for NSW Health Commission, 'Location and Condition of Previous Tip Sites', 18 March 1977.

6   Interview, Father John Boyle, 30 October 2004.

7   Boyle v James Hardie & Coy Pty Ltd, NSWDDT 186 1988.

8   Attrill op. cit.

9   D. Gleadall, affidavit, 12 October 2004, Gleadall v Amaca Pty Limited (formerly James Hardie and Coy Pty Limited), NSWDDT 140 2004.

10 Anthony Cini, affidavit, 17 February 2003, Cini v Amaca Pty Limited (formerly James Hardie & Coy Pty Ltd), NSWDDT 376 2002.

11 ibid.

12 Lawrence Buttigieg, affidavit, 21 October 2003, Lawrence Buttigieg v Amaca Pty Limited, NSWDDT 158 2003.

13 Interview, Meredith Hellicar, op. cit.

14 Sandy Bond and David Cook, 'Residents' perceptions towards asbestos contamination of land and its impact on residential property values', *Pacific Rim Property Research Journal*, vol. 10, no. 3, 2003.

15 Interview, Cliff McCord, 19 February 2008.

16 Manukau Council, news release, 'Largest asbestos removal project nears completion', 23 May 2003; Alan Perrott, 'Asbestos: The killer in the soil', *New Zealand Herald*, 20 June 2001; Interview Cliff McCord, op. cit.

17 Lisa-Jane Loch, 'The NZ Situation', confidential report for JHI NV, 14 April 1999.

18 ibid.

19 Lisa-Jane Loch, 'Minimising the impact on New Zealand activities', James Hardie corporate affairs, 20 August 1999.

20 Peter Shafron, JH board paper, 'Asbestos', 4 February 2000, Part B: 'Possible New Zealand Polluter Pays Laws', Jackson, op. cit., p. 4, exhibit 22.

21 Interview, Joan Sandilands, 21 May 2005.

22 Open letter from John Reid, chairman, James Hardie Asbestos Limited, 28 March 1979.

23 'Asbestos dumps are safe', News, 21 March 1979, quoting the NSW health minister Frank Stewart. Stewart also answered a question in the NSW Legislative Assembly, 21 March 1979, LA 119.

24 Letter from H. Bailey, regional health inspector, NSW Health Commission, Western Metropolitan Health Region, to Parramatta City Council, 4 July 1979; letter from Sam Khoudair, scientific officer, industrial hygiene branch, NSW Division of Occupational Health, to Charles Russell, James Hardie, 12 April 1979.

25 In a letter to the council on 3 May 1979, Hardie's acting NSW manager, C. S. Hoyes commented: 'Emotive statements on this issue have appeared in the news media ... it is our opinion there are no hazards to the health of the people of Parramatta ... Nevertheless ... because we are keen to maintain the good relations we have enjoyed with Parramatta City Council, we are prepared to make, without prejudice, a contribution ...'.

26 Earlier Hardie had a Dust Control Committee, but most of its functions were absorbed into what was then called the Environmental Control Committee.

27 Minutes of the inaugural meeting of the James Hardie Environmental Control Committee, 22 January 1971.

28 Interview, Tony Abela, with Mick O'Donnell, 7.30 Report, ABC TV, 28 October 2004.

29 JH Environmental Control Committee minutes, 29 March 1971, p. 3.

30 Robert Taylor, 'Main Roads knew of Asbestos Link in 2000',
   West Australian, 8 October 2002; Robert Taylor and Steve
   Pennells, 'Asbestos exposed: Two-year battle to clean up
   Burswood rail link', *West Australian*, 8 October 2002.
31 ibid.
32 Michael Southwell, 'Risk to public's health discounted',
   West Australian, 8 October 2002.
33 Interview, Peter Macdonald, 11 November 2003.
34 Under the heading 'Waste Disposal' in the minutes of JH's
   Environmental Control Committee meeting on 12 April 1972, is
   noted: 'Mr Cohen believed that it was here that our problems
   would next arise ...'
35 Interview, John Harris, 19 November 2005.
36 Terrence Miller, affidavit, 30 May 2002, Terrence Miller v Amaca
   Pty Limited (formerly James Hardie & Coy Pty Limited),
   NSWDDT 120 2001.
37 Interview, Terry Miller, with Mike Sexton, *7.30 Report*, ABC TV,
   29 October 2004.
38 JH Environmental Control Committee minutes, 11 December 1974.
39 Once digging has commenced, of course, the dust is disturbed and
   the danger is present. Even Parramatta Council building
   inspectors were exposed to potentially lethal levels of dust in later
   years when they inspected some of the excavation sites in the
   Camellia and Rosehill area where Hardie had dumped tonnes of
   its waste. Roberts Tomlinson, affidavit, 10 February 2006,
   NSWDDT 5142 2005.
40 Confidential JH memo from Neil Gilbert, manager, research and
   development, to director Frank Page, 'Employee Medical Service
   — Wunderlich Ltd', 6 February 1968.
41 ibid.
42 Confidential JH memo from Neil Gilbert to D. S. Waters,
   'Industrial Hygiene', 2 December 1968.
43 It was not the only recommendation by Gilbert that was ignored.
   In July 1968, he insisted that asbestos waste collected in the dust

hoppers must no longer be used as fill, either by employees or
around the Hardie factories (notes from factory managers'
conference 17–21 June 1968); more than a year later he was still
asking Hardie to set out a policy on the disposal of collected dust
(confidential JH memo to Warwick Lane, then manager of the
largest plant at Sydney's Camellia, 'Industrial Hygiene', 26 August
1969). Anecdotal evidence suggests waste continued to be
dumped in the area around Hardie's South Australian factories
until the mid 1980s.

44 Interview, John Harris, op. cit.

45 Phillip Hammond, 'Asbestos warning to farmers', *Courier-Mail*,
    29 October 1992.

46 Paul Paroz v James Hardie & Coy Pty Ltd, NSWDDT 101 1997.

47 From the files of Dr David Kilpatrick, Kilpatrick and Associates,
    and Slater & Gordon, solicitors.

48 W. Musk, N. Olsen, A. Reid, T. Threfall and N. de Klerk,
    'Asbestos-related disease from recycled hessian superphosphate
    bags in rural Western Australia', *Australian and New Zealand
    Journal of Public Health* 2006; vol. 30, pp. 312–313.

49 Kevin Barrett, statement, 25 March 2004 (in the author's
    possession).

50 Marnie McKimmie, 'New asbestos threat in thousands of WA
    homes', *West Australian*, 24 March 2006.

51 News release, WA Department of Health, 10 July 2006.

52 Interview, Mike Lindsay, WA Department of Health, 16 March
    2006.

53 Interview, Robert Vojakovic, 17 March 2006.

54 John Downes, statement, John Downes v Amaca Pty Limited
    (formerly James Hardie & Coy Pty Ltd), NSWDDT 6097 2006.

55 JH Environmental Control Committee minutes, 6 October 1971.

56 Interview, Meredith Hellicar, op. cit.

57 John B. Reid, *Commonsense Corporate Governance*, Australian
    Institute of Company Directors, Sydney, 2002, p. 107.

BIBLIOGRAPHY

Australian Parliament House of Representatives Standing Committee on Aboriginal Affairs and G. L. Hand, *The effects of asbestos mining on the Baryulgil community: report of the House of Representatives Standing Committee on Aboriginal Affairs*, AGPS, Canberra, 1984.

Margo Beasley, *The Missos*, Allen & Unwin, Sydney, 1996.

Michael Bowker, *Fatal Deception*, Rodale, Los Angeles, 2003.

Paul Brodeur, *Expendable Americans*, Viking Press, New York, 1974.

Paul Brodeur, *Outrageous Misconduct: The Asbestos Industry on Trial*, Pantheon, New York, 1985.

Ian Campbell, *History of James Hardie and Coy. Pty. Limited, 1888 to 1966*, James Hardie & Coy. Pty Ltd, Sydney, 1967

Janet Cannon, *Yugilbar, 1949–1999*, Hardie Grant Publishing, Melbourne, 1999.

Michael Cannon, *That Disreputable Firm: The Inside Story of Slater & Gordon*, Melbourne University Press, 1998.

Brian Carroll, *'A Very Good Business': one hundred years of James Hardie Industries Limited 1888-1988*, James Hardie Industries Limited, Sydney, 1987.

Barry Castleman, *Asbestos: Medical and Legal Aspects*, 5th edition, Aspen Publishers, New York, 2005.

Pauline Clayton, *Remember Newstead*, James Hardie & Coy Pty Ltd, Brisbane, 1992.

George Farwell, *Squatter's Castle: The Story of a Pastoral Dynasty*, Landsowne Press, Melbourne, 1973.

Rosey Golds, *A History of the Dust Diseases Board, 1927–2007*, DDB, Sydney, 2008.

Gideon Haigh, *Asbestos House*, Scribe, Melbourne, 2006.

Timothy Hall, *The Ugly Face of Australian Business*, Harper & Row, Sydney, 1980.

Ben Hills, *Blue Murder*, Sun Books, Sydney, 1989.

David Jackson, *Report of the Special Commission of Inquiry into the Medical Research and Compensation Foundation*, Department of the Premier and Cabinet, NSW Government, September 2004.

Suzanne LeBlanc, *Cassiar, A Jewel in the Wilderness*, Caitlin Press, Toronto, Canada, 2003.

Ray Gietzelt, *Worth Fighting For*, Federation Press, Sydney, 2004.

Jock McCulloch, *Asbestos: Its Human Cost*, University of Queensland Press, St Lucia, 1986.

Jock McCulloch, *Asbestos Blues: Labour, Capital, Physicians and the State in South Africa*, Indiana University Press, 2002.

Jock McCulloch and Geoffrey Tweedale, *Defending the Indefensible*, Oxford University Press, 2008.

Matt Peacock, *Asbestos: Work as a Health Hazard*, ABC (with Hodder & Stoughton), Sydney, 1978.

John B. Reid, *Commonsense Corporate Governance*, Australian Institute of Company Directors, Sydney, 2002.

Elizabeth Thurbon, *Climbing Out of the Big Black Asbestos Hole*, Canberra, 2004.

Geoffrey Tweedale, *Magic Mineral to Killer Dust*, Oxford University Press, 2000.

George Wragg, *The Asbestos Timebomb*, Catalyst Press, Sydney, 1995.

AUTHOR INTERVIEWS

Mostly face-to-face interviews were conducted for this book, in many cases multiple interviews were undertaken. The interviews were generally recorded, then transcribed. For ease of reference the endnotes record only the date of the first interview, unless it is clearly inappropriate.

Interviewees: Kevin Baker, Bernie Banton, Karen Banton, Paul Bastian, Phillip Batson, Greg Baxter, Harry Black, Thady Blundell, Father John Boyle, David Britton, Joe Calibrese, Doug Cameron, George Campbell, Bob Carr, Barry Castleman, Angus Cave, Greg Combet, Dennis Cooper, Michael Costa, Phil Davis, Eileen Day, Bob Debus, John Della Bosca, Jim Donovan, Allen Drew, Sir Llewellyn Edwards, Laurie Ferguson, Stuart Fist, Vic Fitzgerald, Ken Fowlie, Eva Francis, Brett Freeburn, Bill Frew, Armando Gardiman, Ray Geitzelt, Neil Gilbert, Jenny Gleadall, John Gordon, Peter Gordon, Gary Gray, Sir James Hardy, Meredith Hellicar, Professor Douglas Henderson, Ben Hills, Ron Hinton, Ray Hogan, Doug Howitt, Ian Hutchinson, Jon Isaacs, Dr Anthony Johnston, Dr Ray Jones, Laurie Kazaan-Allen, James Kelso, Margaret Kent, David Kilpatrick, Mark Knight, Ian Kortlang, Warwick Lane, Geoff Lansley, Chris Lawrence, Dr Julian Lee, Dr James Leigh, David Leitch, Peter MacDonald,

Warwick Macdonald, Lilly Madden, Gersh Major, Bill Mansfield, Steve Marsellos, Cliff McCord, Jock McCulloch, Richard Meeran, David Miller, Terry Miller, Nancy Milne, Professor William Musk, Ian Mutton, Sir Julian Peto, Jörg Probst, Professor Bruce Robinson, Barry Robson, Jack Rush QC, Peter Russell, Fred Sandilands, Joan Sandilands, Kate Sandilands, David Say, Adam Searle, Tanya Segelov, Eileen Shanahan, Jeff Shaw, John Sheahan SC, Chris Shields, Michael Slattery QC, Mick Smith, Ella Sweeney, David Taylor, Elizabeth Thurbon, Robert Vojakovic, Rose Marie Vojakovic, Frank Walker, John Wells, Roger Wilkins, George Wright and Vic Zammit.

ASBESTOS SUPPORT GROUPS

There are a number of support groups throughout Australia that provide information and assistance for people with asbestos-related diseases. Most groups offer counselling, advice and referrals to trusted doctors and lawyers, and above all, an opportunity to talk to those who have shared similar experiences. Some groups are entirely voluntary and have limited resources; others, like the ADS in Perth, are highly professional organisations and provide a wide range of services. Many groups are partially funded by asbestos litigation firms and in some cases compete for membership; they are nonetheless far more independent than alternative services offered by governments or former asbestos companies.

Asbestos Diseases Society of Australia, Inc (ADS)
219 Main Street
(PO Box 1394)
Osborne Park WA 6017
08 9344 4077
1800 646 690
adsinc@iinet.net.au
www.asbestosdiseases.org.au

Asbestos Diseases Foundation of Australia, Inc. (ADFA)
133–137 Parramatta Road
Granville NSW 2142
02 9637 8759
1800 006 196
info@adfa.org.au
www.adfa.org.au

Asbestoswise
1st Floor
247–251 Flinders Lane
Melbourne VIC 3000
03 9654 9555
info@asbestoswise.com.au
www.asbestoswise.com.au

Gippsland Asbestos Related Diseases Support, Inc. (GARDS)
41 Monash Road
Newborough VIC 3825
03 5127 7744
0407 274 173
info@gards.org
www.gards.org

Asbestos Victims Association (SA), Inc. (AVA)
Level 3, 60 Waymouth Street
Adelaide SA 5000
08 8212 6008
1800 665 395
info@avasa.asn.au
www.avasa.asn.au

Asbestos Diseases Society of South Australia (ADSSA)
30–40 Hurtle Square,
Adelaide SA 5000
08 8359 2423
1800 157 540
admin@adssa-inc.com.au
www.adssa-inc.com.au

**Queensland Asbestos Related Diseases
Support Society, Inc. (QARDSS)**
16 Campbell Street
Bowen Hills, Brisbane QLD 4006
07 3252 7852
1800 776 412
asbestoshelp@westnet.com.au
www.asbestos-disease.com.au

Bernie Banton Foundation
PO Box 98
Tyabb VIC 3913
0412 830 485
info@berniebanton.com.au
www.berniebanton.com.au

The Asbestos Free Tasmania Foundation
180–184 Collin Street
Hobart TAS 7000
03 6233 7950
info@asbestosfreetasmania.org.au
www.asbestosfreetasmania.org.au

Links to overseas groups and useful information is also available from:

The International Ban Asbestos Secretariat (IBAS)
www.ibas.btinternet.co.uk

ACKNOWLEDGEMENTS

My heartfelt thanks go to my family, who have put up with me while I wrote this book, especially my partner Kerry, my daughter Savannah and stepdaughters Cato and Micky, and my son Dylan. A special thanks to Liz Fell, who was amazingly generous with her time, giving advice, encouragement and criticism, and without whom the book probably never would have been finished.

At ABC Books, Susan Morris-Yates never wavered in her support, Ali Lavau was an effortless editor, and Brigitta Doyle steered the book through the new relationship with HarperCollins, where I received total support. My cousin Carolyn, my sister Deb and her husband Tim provided a safe haven in which to write (where they put up with my raves about the subject — and I dodged the occasional snake); my ABC colleagues also put up with similar distractions and many lent a hand. The librarians at the ABC, especially Keryn Kellaway and Cathy Beale, the Mitchell Library at the State Library of NSW, Mary Wearin at the Dust Diseases Tribunal, and particularly Jörg Probst at Turner Freeman have been invaluable.

Special thanks are due to Turner Freeman's Armando Gardiman and Tanya Segelov for their time and encouragement.

I am grateful for the help of the many other lawyers, academics unionists and other activists who have generously assisted: Robert and Rose Marie Vojakovic, Jack Rush, Peter Gordon and his colleagues at Slater & Gordon, especially Ken Fowlie, Michael Magazanik, Tim Hammond and Ben Phi, Ian Hutchinson, Jim Leigh, Jock McCulloch, Laurie-Kazaan Allen, Barry Castleman, Geoffrey Tweedale, David Taylor, Lilly Madden, Jane Singleton and many others too numerous to mention here. There are bound to be people I wanted to thank, but have stupidly forgotten in my last-minute rush. Please accept my apology in advance. And a big thanks to all the others who urged me on, no matter how long it took!

Index

JH = James Hardie

Matt Peacock is an award-winning TV and radio journalist who has filled many of the ABC's key correspondent positions, including chief political correspondent for ABC current affairs radio Canberra, as well as Washington, New York and London correspondent.

After working on Australia's first TV current affairs show, *This Day Tonight*, he joined the ABC Science Unit where he made a pioneering series on the Australian asbestos industry in 1977.

He revisited the subject 30 years later, as a reporter with ABC TV's *7.30 Report*, then writing this book.

His role as an investigative reporter has been dramatised in an ABC1 mini-series based on this book, *Devil's Dust*, produced by Fremantle Media.

Matt has written for many newspapers and magazines and has authored and contributed to several books. He is currently an Adjunct Professor with Sydney's University of Technology (UTS).

He lives in Sydney.